Advances in Fluidization Engineering

AIChE Symposium Series

Liang-Shih Fan, Volume Editor

276   *Volume 86, 1990*

AMERICAN INSTITUTE OF CHEMICAL ENGINEERS

Advances in Fluidization Engineering

Elmer L. Gaden, Jr., Series Editor

Liang-Shih Fan, Volume Editor

Maura Mullen, Manager Publications Production

K. Alia	T.C. Ho	P. Nicoletti
I. Alkan	J.R. Hopper	Joseph J. Perona
F. Bayat	C.C. Huang	N.S. Rao
F. Bentahar	L.B. Hutchinson	S.C. Saxena
R.H. Birk	S. Iwata	S. Shukla
G.A. Camp	T.M. Knowlton	D.C. Tian
I. Chan	B. Kocatulum	George H. Webster
J.M. Chen	H.O. Kono	Keith D. Wisecarver
L.V. Dullea	J.F. Large	Kuo-Ying Amanda Wu
Arunava Dutta	E.K. Levy	I. Yamada
G.E. Fasching	T. Matsuda	A. Yamamoto
J.G. Findlay	E. Mizutani	P.L. Yue
J.S. Halow	Y. Molodtsof	N. Zarifis
T. Haruta	S. Mori	S.J. Zhou

Inquiries regarding the publication of Symposium Series Volumes should be directed to
Dr. Elmer L. Gaden, Jr., Series Editor,
University of Virginia, Department of Chemical Engineering,
Thornton Hall, Charlottesville, Virginia 22903-2442.   FAX (804) 924-6270.

AIChE Symposium Series

1990

Published by
American Institute of Chemical Engineers

Copyright 1990

American Institute of Chemical Engineers
345 East 47 Street, New York, N.Y. 10017

AIChE shall not be responsible for statements or opinions advanced in papers or printed in its publications.

Library of Congress Cataloging-in-Publication Data

Advances in fluidization engineering / Liang-Shih, editor.
    p. cm. — (AIChE symposium series ; no. 276, v. 86, 1990)
  Contains 14 papers presented at the AIChE Annual Meeting, held San Francisco, Nov. 5-10, 1989.
  Includes index.
  ISBN 0-8169-0488-X
  1. Fluidization—Congresses.   I. Fan, Liang-Shih. II. American Institute of Chemical Engineers. Meeting (1989 : San Francisco, Calif.)   III. Series: AIChE symposium series ; no. 276.
TP156.F65A25 1990
660'.284292—dc20
                                                                          90-837
                                                                            CIP

    Authorization to photocopy items for internal or personal use, or the internal or personal use of specific clients, is granted by AIChE for libraries and other users registered with the Copyright Clearance Center (CCC) Transactional Reporting Service, provided that the $2.00 fee per copy is paid directly to CCC, 21 Congress St., Salem, MA 01970. This consent does not extend to copying for general distribution, for advertising, or promotional purposes, for inclusion in a publication, or for resale.

    Articles published before 1978 are subject to the same copyright conditions and the fee is $2.50 for each article. AIChE Symposium Series fee codee: 0065-8812/90 $2.50.

FOREWORD

Fluidization and Fluid particle systems have been commonly encountered in industries involving, e.g., physical, chemical and petrochemical operations. In recent years, engineering developments in biotechnology and advanced materials have called for an extensive employment of these systems. Much, however, remains to be understood in these complex multiphase systems.

The AIChE Annual Meeting has traditionally been an important national and international forum for researchers to discuss and present their new findings. It also provides a forum for those who are interested in the field to gain state-of-the-art knowledge.

This volume of the AIChE Symposium Series encompasses 14 papers selected from those presented in the five sessions at the Annual Meeting held in San Francisco, November 5-10, 1989. These papers cover a wide spectrum of important fundamental and application topics in fluidization and fluid particle systems. Such topics include hydrodynamics and heat transfer in a gas- or liquid-fluidized bed, gas and solids flow characteristics, gas distributor design, heat exchanger tube erosion, fine powder fluidization and means for improving fine powder fluidization quality, measurement techniques, and chemical and biochemiocal reactor applications.

I would like to express my appreciation to the Chairmen and Co-Chairmen of the sessions for their excellent efforts in the program planning, and to the reviewers who provided valuable comments on the papers included in this volume.

Liang-Shih Fan, *editor*
The Ohio State University
Columbus, OH 43210

Dedicated to the memory of
Professor L.S. (Ming) Leung
a pioneer in the field of
fluidization and fluid particle systems.

CONTENTS

FOREWORD iii

ANALYSIS OF THE EFFECTS OF BUBBLE COALESCENCE ON TUBE WASTAGE .... F. Bayat, E. Levy and I. Alkan 1

HEAT TRANSFER TO VERTICALLY FLOWING DILUTE AND DENSE PHASE
   GAS-SOLIDS SUSPENSIONS .... F. Bentahar, Y. Molodtsof, J.F. Large and K. Alia 10

DESIGN OF AN AGGLOMERATION RESISTANT GAS DISTRIBUTOR .... R.H. Birk, G.A. Camp and L.B. Hutchinson 16

A COMPARATIVE EVALUATION OF NEGATIVELY AND POSITIVELY CHARGED
   SUBMICRON PARTICLES AS FLOW CONDITIONERS FOR A COHESIVE POWDER .... Arunava Dutta and L.V. Dullea 26

PRELIMINARY CAPACITANCE IMAGING EXPERIMENTS OF A FLUIDIZED BED ... J.S. Halow, G.E. Fasching and P. Nicoletti 41

METAL CAPTURE DURING FLUIDIZED BED INCINERATION OF SOLID WASTES
   .... T.C. Ho, J.M. Chen, S. Shukla and J.R. Hopper 51

CONTINUOUS DEPRESSURIZATION OF SOLIDS USING A RESTRICTED
   PIPE DISCHARGE SYSTEM .... T.M. Knowlton, J.G. Findlay and I. Chan 61

AGGLOMERATION CLUSTER FORMATION OF FINE POWDERS
   IN GAS-SOLID TWO PHASE FLOW .... H.O. Kono, T. Matsuda, C.C. Huang, and D.C. Tian 72

ANALYSIS OF GAS MOTION AT THE SURFACE OF A FLUIDIZED BED
   DUE TO BUBBLE ERUPTIONS .... E.K. Levy and B. Kocatulum 78

VIBRO-FLUIDIZATION OF GROP-C PARTICLES AND ITS INDUSTRIAL
   APPLICATION .... S. Mori, A. Yamamoto, S. Iwata, T. Haruta, I. Yamada, and E. Mizutani 88

FLUIDIZATION REGIME DELINEATION IN GAS-FLUIDIZED BEDS .... S.C. Saxena, N.S. Rao and S.J. Zhou 95

THE EFFECT OF TAPER ANGLE ON THE HYDRODYNAMICS OF A TAPERED
   LIQUID-SOLID FLUIDIZED BED .... George H. Webster and Joseph J. Perona 104

BIOLOGICAL PHENOL DEGRADATION IN A COUNTERCURRENT
   THREE-PHASE FLUIDIZED BED USING A NOVEL CELL
   IMMOBILIZATION TECHNIQUE .... Kuo-Ying Amanda Wu and Keith D. Wisecarver 113

CATALYTIC DEHYDRATION OF ETHANOL IN A FLUIDIZED BED REACTOR .... P.L. Yue and N. Zarifis 119

INDEX 126

ANALYSIS OF THE EFFECTS OF BUBBLE COALESCENCE ON TUBE WASTAGE

F. Bayat, E.K. Levy and I. Alkan ■ Energy Research Center, Lehigh University, Bethlehem, PA 18015

Recent studies have shown that the erosion on the bottom half of a horizontal tube in a bubbling fluidized bed is caused by the upward impact of bubble wakes on the tubes. These studies have also shown that the high velocity wakes formed by the vertical coalescence of two bubbles cause particularly high rates of erosion. This paper describes the development of a theoretical model which first computes bubbling frequency and bubble coalescence rates. The model uses this information along with empirical information on the rates of erosion caused by individual bubbles to determine overall rates of erosion of tubes in the bottom row of the bundle in a fluidized bed. Sample results from a model, presented for the TVA 20 MW FBC pilot plant, illustrate typical variations of rate of erosion with changes in the distance between the bottom row of tubes in the tube bundle and the distributor.

INTRODUCTION

The use of fluidized bed combustion (FBC) for industrial and utility boilers has grown rapidly in recent years. FBC technology has met most expectations from the points of view of both combustion performance and pollution control. However, in some units, maintenance problems have arisen due to unexpectedly high rates of metal loss from the surfaces of heat exchanger tubes.

In the case of horizontal tubes located at the bottom of a tube bundle in a bubbling fluidized bed, the greatest rates of erosion are on the bottom halves of the tubes and these erosion rates are caused by the impact of bubble wakes on the tubes. In controlled experiments in a room temperature fluidized bed, Levy and Bayat (1) showed a direct correlation between rate of thinning of the tube wall and the velocity of the wake of the impacting bubble. They also found that when a pair of bubbles coalesces beneath the tube, a high velocity jet of bed material rises up out of the wake of the trailing bubble. The velocity of this wake jet was found to be approximately 2.3 times the velocity of the wake of a single bubble of the same size. The experiments of Levy and Bayat confirmed that because of their higher wake velocities, bubbles in the process of coalescing as they strike a tube are more damaging to tube surfaces than single bubbles of the same size.

Rathbone, et al. (2) measured the particle velocities and stresses on a tube surface during bubble impact and found that the highest stresses occur when the bubble wake strikes the tube. Zhu, et al. (3) performed flow visualization studies and tube wastage measurements and also confirmed the importance of the bubble wake in causing erosion. Both Rathbone, et al. and Zhu, et al. identified bubble coalescence as being a very damaging type of bubble/tube interaction.

The identification of the impact of the bubble wake as the mechanism for erosion damage makes it possible to develop predictive models that link the fluid mechanics of the bed to rates of erosion. While the previously described studies all show the effects of individual bubble/tube interactions and the relative importance of single bubbles and coalescing bubble pairs (double bubbles) in causing erosion, they do not provide information on the relationship between bed design and operating conditions to erosion rate. This paper describes the development of a theoretical model which computes bubbling frequency and bubble coalescence rates and then uses this information with empirical information on the rates of erosion caused by individual bubbles to determine overall rates of erosion of tubes in a fluidized bed. The model is restricted to the analysis of erosion rates of horizontal tubes in the bottom row of a tube bundle in a bubbling fluidized bed.

BUBBLE COALESCENCE MODEL

Previous Work

Because of the importance of bubbles in fluidized beds, the motion of interacting bubbles has been the subject of numerous studies. Among these, Harrison and Leung (4) studied the coalescence of an isolated bubble pair in vertical alignment in a small particle fluidized bed. It was observed that if the nose of one bubble is within the wake of another, then the bubble velocity of the trailing bubble is increased by an amount approximately equal to the velocity of the leading bubble and coalescence takes place. According to the observations of Rowe (5) and Toei and Matsuno (6), when two obliquely aligned bubbles interact, the front bubble rises almost vertically while the rear bubble moves laterally to a position behind a leading bubble. The rear bubble then accelerates vertically so that coalescence occurs with the bubble alignment almost vertical.

One of the early theoretical studies on bubble coalescence was carried out by Lin (7) for vertically aligned, two-dimensional bubble pairs. Gabor (8) studied chains of equally spaced two- and three-dimensional bubbles. The most detailed theoretical work on bubble coalescence to predict the behavior of any number of bubbles in a freely bubbling bed was done by Clift and Grace (9), Nguyen, et al. (10) and Farrokhalee and Clift (11).

Description of Model

The work of Clift and Grace was used as the basis for the bubble coalescence model used in the present paper. The motion of the bubble pair shown in Figure 1 was analyzed subject to the following assumptions:

- The particulate phase behaves as an incompressible, irrotational fluid.

- A spherical boundary envelops each bubble and its wake. The boundary moves with the bubble so that the motion outside this boundary is governed by potential flow theory.

- The velocity field of each bubble is represented by a single doublet positioned at its center and moving with zero velocity relative to the bubble.

The velocity of a bubble is approximated by adding to its rise velocity in isolation, the velocity which the emulsion phase would have at the position of the bubble nose, if the bubble were absent.

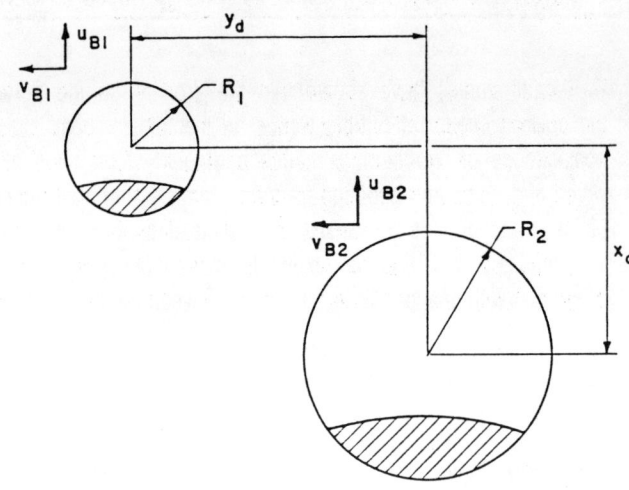

Figure 1. Sketch of a pair of bubbles interacting in two dimensions.

Results of previous investigators have shown that during an interaction between bubbles, the trailing bubble has only a very small effect on the motion of the leading bubble. To simplify the analysis, the effect of the trailing bubble on the motion of the leading bubble is neglected in the present investigation.

With these assumptions, the velocity equations for a pair of spherical bubbles interacting in two dimensions take the forms:

$$U_{B1} = U_{A1}$$

$$V_{B1} = 0$$

$$U_{B2} = U_{A2} + q_{21}$$

$$V_{B2} = p_{21}$$

Where "1" and "2" refer to the leading and trailing bubbles and where q, p are the components of the particulate phase velocity at the nose of the trailing bubble due to the motion of the leading bubble.

The isolated rise velocity of the leading bubble is obtained from

$$U_{A1} = 0.711\sqrt{gD}$$

In addition, from potential flow theory, the velocity components p_{21} and q_{21} are

$$q_{21} = \frac{[2(x_d + R_2)^2 - y_d^2] R_1^3 U_{B1}}{2[(x_d + R_2)^2 + y_d^2]^{5/2}}$$

$$p_{21} = \frac{3(x_d + R_2) y_d R_1^3 U_{B1}}{2[(x_d + R_2)^2 + (y_d)^2]^{5/2}}$$

When the trailing bubble enters the wake of the leading bubble, the leading bubble is still not affected, but now the equations take the following form:

$$U_{B1} = U_{A1}$$

$$U_{B2} = U_{A2} + U_{B1}$$

$$V_{B2} = V_{B1} = 0$$

In addition, the following assumptions were made for that part of the coalescence event which occurs after the nose of the trailing bubble penetrates the wake of the leading bubble.

- The nose of Bubble 2 enters the wake of Bubble 1 close to the vertical line through the center of Bubble 2.

- Coalescence is complete after a characteristic time interval $\Delta t_c = 1.8 R_1/U_{B1}$, beginning with the initiation of nose-wake entry. (See Ref. 9)

- The bubble volumes are additive during coalescence.

Once the instantaneous velocities are known, the bubble trajectories are obtained by numerical integration of the equations

$$\frac{dx_i}{dt} = U_{Bi}$$

$$\frac{dy_i}{dt} = V_{Bi}$$

Experimental Verification

Flow visualization experiments were carried out in a 775 mm x 775 mm cross section room temperature bed to validate the model and confirm the assumptions. The experiments were performed with both 0.8 mm and 1.3 mm sand particles with the bed at incipiently fluidized conditions. Bubble injectors, located adjacent to a transparent side wall, were used to create bubbles of controlled size and initial positions. These bubbles, which were formed near the distributor and at a side wall, flowed upward along the wall such that they were visible from outside the bed. A high speed video system was used to record the bubble motions as the bubbles interacted.

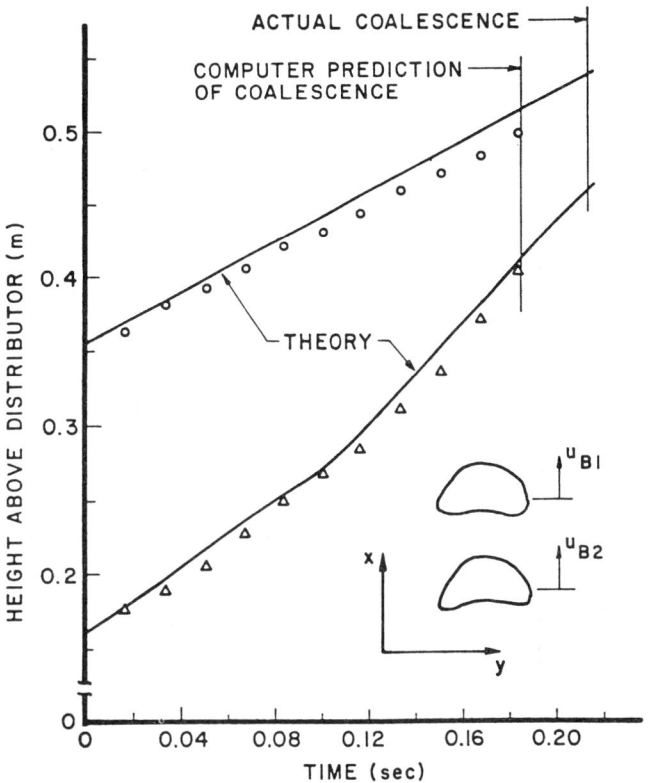

Figure 2. Bubble nose positions versus time for a vertically aligned bubble pair in a fluidized bed near minimum fluidization conditions (d_p = 0.8 mm).

Figure 2 shows the nose positions versus time for a pair of vertically aligned bubbles of equal size. The solid lines give the results of the computer model, while the symbols give the measured values. The leading bubble moves at constant velocity at its isolation rise velocity

$$U_{A1} = 0.711\sqrt{gD}$$

while the trailing bubble is steadily accelerated. At 0.11 seconds, the nose of the trailing bubble enters the wake of the leading bubble, causing an increase in velocity and a change in the slope of the trajectory of the trailing bubble. The time of actual coalescence corresponds to the point at which the trailing bubble wake starts to erupt. The computer model predicts this to occur at an interval of time $\Delta t_c = 1.8\, R_1/U_{B1}$ after the penetration of the wake of the leading bubble.

A similar bubble nose position versus time relationship is presented in Figures 3 and 4 for an obliquely aligned bubble pair.

At about 0.17 seconds, the trailing bubble becomes vertically aligned with the leading bubble and from here on they move together until the point of coalescence.

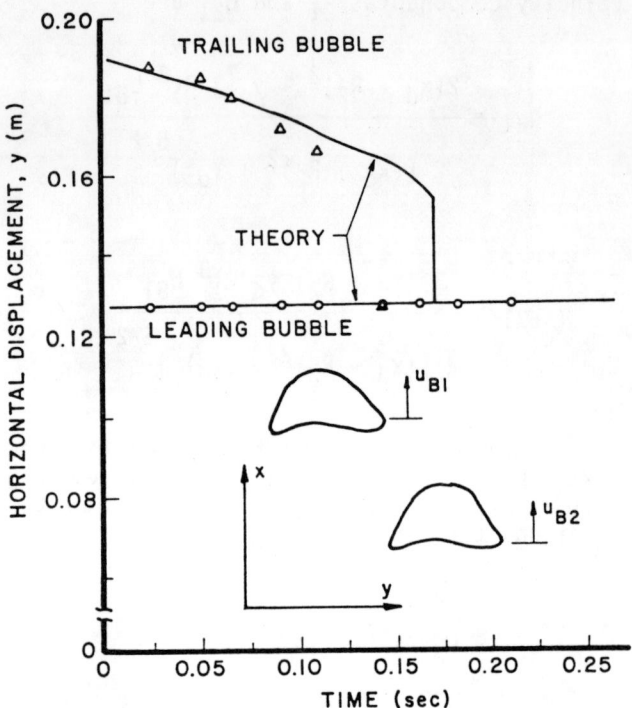

Figure 4. Variation of horizontal distance between the leading and trailing bubbles in a fluidized bed near minimum fluidization conditions (d_p = 0.8 mm).

Figure 5 shows the interactions of a pair of equal size bubbles which are aligned horizontally. These rise together, moving side by side without coalescing.

The results shown in Figures 2 to 5 are typical of many bubbles studied in the experiments. Excellent agreement was obtained between the theory and experiments and the assumption of negligible effect of the trailing bubble on the motion of the leading bubble was verified. Similar conclusions were also drawn by Harrison and Leung (4).

Extension to Three-Dimensional Freely Bubbling Case

The basic approach described above was generalized so it could be used to model bubble motion within a freely bubbling three-dimensional bed containing N bubbles. The

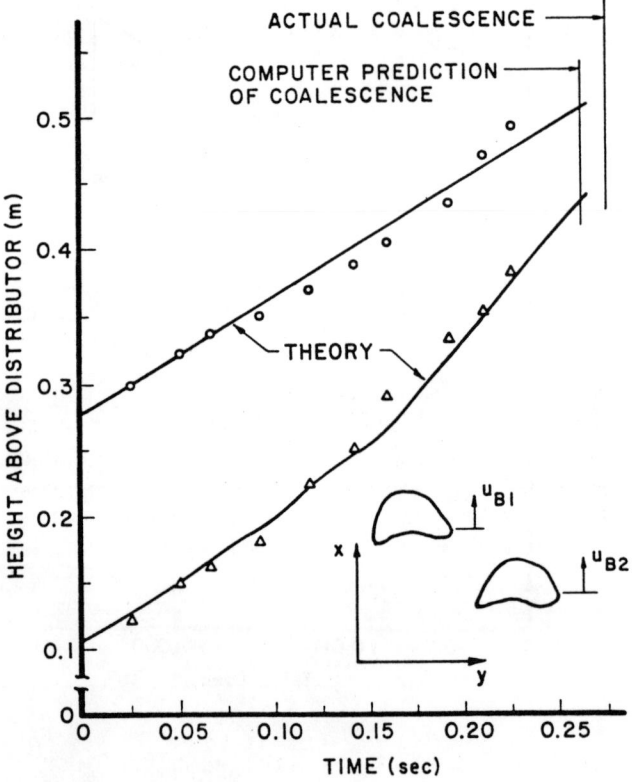

Figure 3. Bubble nose positions versus time for an obliquely aligned bubble pair in a fluidized bed near minimum fluidization conditions (d_p = 0.8 mm).

equations used to compute the bubble trajectories are similar in concept to those used for the two-dimensional two-bubble problem. A detailed description of these equations and the solution procedure are given in Reference 12.

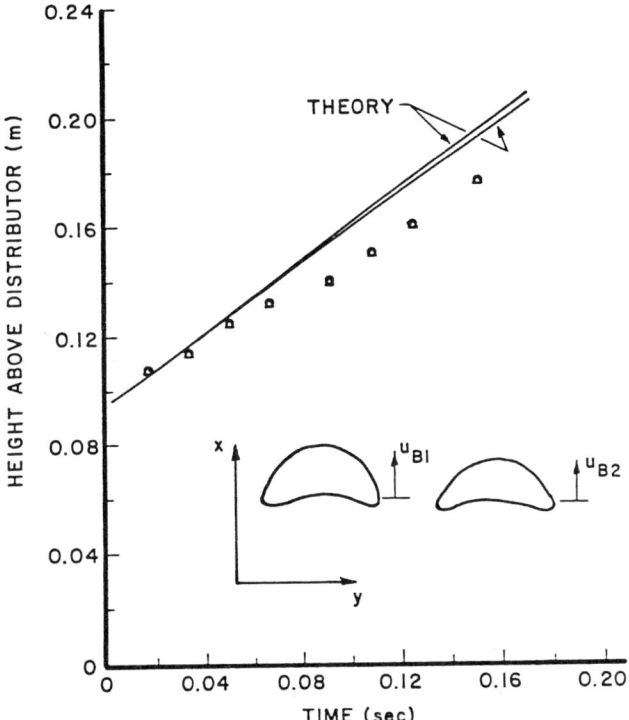

Figure 5. Bubble nose positions versus time for a horizontally aligned bubble pair in a fluidized bed near minimum fluidization conditions (d_p = 0.8 mm).

The freely bubbling case requires information on the size, frequency and spatial distribution of the bubbles at the distributor. The bubble size at the distributor is a function of distributor geometry, and in this case the Mori and Wen correlation (Reference 13) for distributors with discrete gas injection orifices was used to compute initial bubble size.

$$D_o = \frac{1.38}{g^{0.2}} \left[\frac{A_{bed} (U_o - U_{mf})}{N_d} \right]^{0.4}$$

The total, visible bubble flow rate was obtained from the excess gas velocity ($U_o - U_{mf}$) according to

$$Q_B = Y (U_o - U_{mf}) A_{Bed}$$

where for large particles with d_p > 0.7 mm (Geldart type D), $Y \simeq 0.25$ (Reference 14). The frequency of bubble formation at the distributor nozzles was then computed from

$$f = Y \left[\frac{A_{Bed} (U_o - U_{mf})}{N_d} \right] \frac{1}{q_B}$$

At the beginning of an analysis, the sequence of formation of the initial bubbles among the nozzles was chosen randomly, with the same formation frequency used thereafter for all the nozzles.

TUBE EROSION MODEL

The bubble coalescence model described in the previous section was used to generate the information on bubble characteristics needed to compute local tube erosion rates. This information consisted of local values of bubble size, the wake velocities of both single and double bubbles, and the frequencies of occurrence of single and double bubbles at the locations of interest within the bed. In performing the analysis, the region of interest is selected. This region is at the elevation of the bottom row of tubes and is square in cross section. The length of each side of the square region is chosen to be the same as the initial bubble diameter at the distributor. If at the elevation of the tube a bubble nose is located in this area, then that particular bubble is considered to be interacting with the tube.

The bubble size which is used in the erosion calculations is determined by taking the average of all bubbles which pass through the area of interest at that elevation.

$$D_B = \sum_{i=1}^{N} \frac{D_B(X)}{N}$$

The bubble frequency at each reference level is computed from the number of bubble noses which pass through the region of interest divided by the time duration of the simulation

$$f = \frac{\text{(number of interacting bubbles during time } \Delta t)}{\Delta t}$$

The erosion due to the impact of a single bubble is obtained from

$$E_{SB} = a\, V_{SB}^n$$

where in the case of a single bubble, since the bubble nose and wake move at the same speed, V_{SB} is the instantaneous bubble rise velocity. These velocities were obtained directly from the computer simulation.

The erosion due to the impact of the wake of a double bubble on the tube was modeled as

$$E_{DB} = a\, V_{DB}^n$$

with

$$V_{DB} = 2.3\, V^*$$

The quantities V^* and V_{DB} represent single bubble and double bubble wake velocities where the single bubble and trailing double bubble have equal diameters. The value of V^* was taken as an average for a large number of bubbles and was obtained from the simulation by computing velocities of single bubbles midway between their coalescences. The coefficient 2.3 was obtained from the measurements of Reference 1.

The cumulative rate of erosion due to the single and double bubbles was then computed as

$$\frac{E_{total}}{\Delta t} = \left[\frac{\Sigma a\, V_{SB}^n}{\Delta t}\right]_{SB} + \left[\frac{\Sigma a\, V_{DB}^n}{\Delta t}\right]_{DB}$$

RESULTS OF EROSION CALCULATIONS

The coalescence/erosion model was used to simulate bubbling and erosion in the Tennessee Valley Authority (TVA) 20 MW pilot FBC. This bed has a 3.7 m x 7.3 m rectangular cross section, with distributor nozzles which are arranged in a square array. The distance between the nozzle centers is 0.114 m, and the total number of nozzles is 2112. The bed has a minimum fluidization velocity of 0.22 m/S and it is typically operated with a superficial gas velocity of 2.74 m/S. The bed contains a bundle of submerged horizontal tubes, and it has been operated at times with the bottom row of tubes located at a distance of 0.51 m from the distributor and at other times at 0.56 and 0.79 m from the distributor (Ref. 15).

The simulation was carried out for a square region, bounded by a 5 x 5 array of nozzles, representing a region in the center of the distributor far from the walls of the bed. A value of 0.25 was used for the visible bubble flow rate parameter Y and a value n=3 was used for the exponent in the erosion rate equation. Values for n ranging from 2 to 3 are widely cited in the erosion literature as being typical for the materials used for FBC heat exchanger tubes (Ref. 16).

At the beginning of the simulation, the bed does not contain any bubbles. These are introduced at the distributor and they rise upward, eventually filling the bed as quasi-steady bubbling conditions are reached. The results shown here are based on 20 seconds of bed operation, with the first six seconds used to account for this initial transient. Quasi-steady state conditions are assumed to occur when the total number of bubbles in the bed becomes constant (see Figure 6). Figure 7 shows how bubble size changes with respect to elevation in the bed. Bubble growth is due to coalescence, but not all of the bubbles coalesce and as a result, some retain their initial sizes at elevations far above the distributor. This is responsible

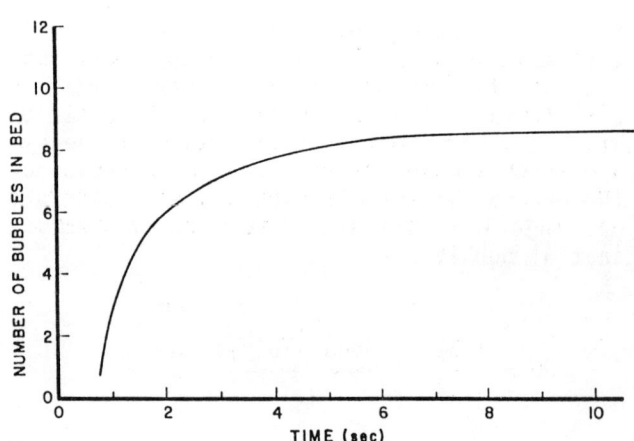

Figure 6. Number of bubbles in the bed versus time for Y = 0.25.

for the relatively large values of standard deviation in bubble size shown in Figure 7. Bubble rise velocities are given as functions of bubble diameter in Figure 8 and are compared with the isolation bubble rise velocity.

Because of coalescence, the number of single bubbles decreases dramatically as the elevation in the bed increases. The single bubble and double bubble frequencies, plotted in Figure 9 show a local peak in double bubble frequency centered at an

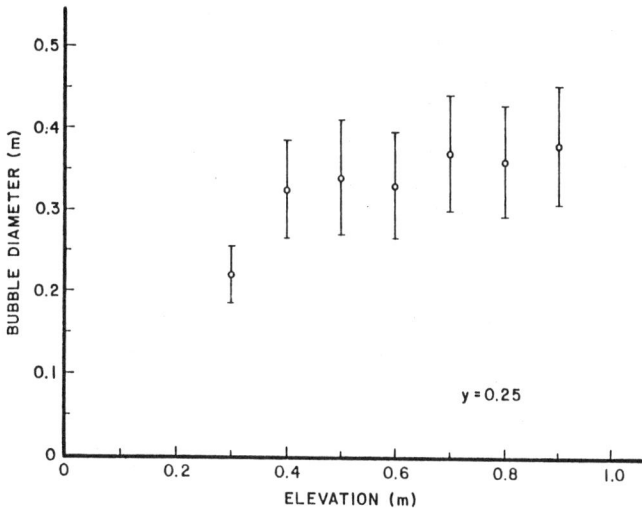

Figure 7. Bubble size variation in the bed for Y = 0.25. These calculated results were obtained from the bubble coalescence model.

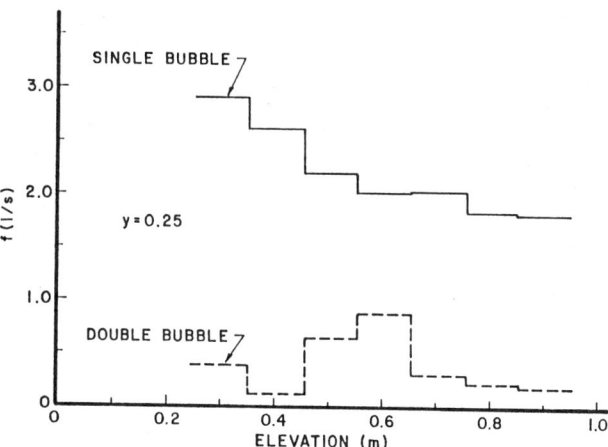

Figure 9. Variation of frequencies of single and double bubbles with distance above the distributor. These results were obtained with the bubble coalescence model.

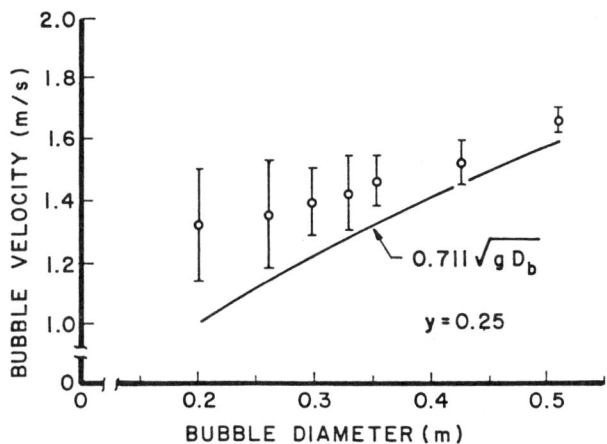

Figure 8. Bubble velocity versus bubble size for Y = 0.25. These results were calculated from the bubble coalescence model.

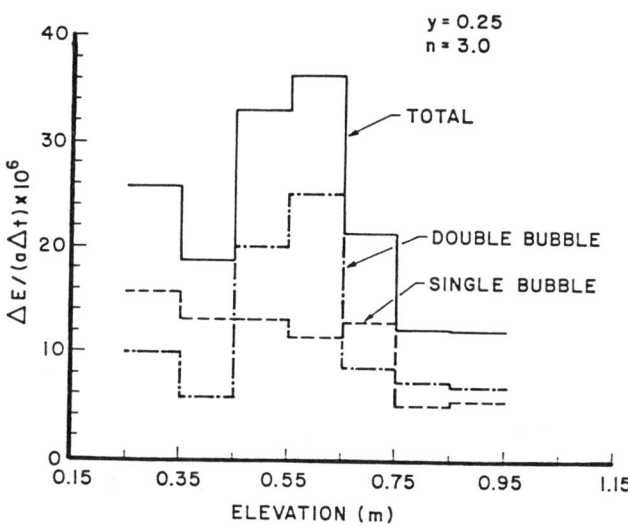

Figure 10. Variation of relative rate of erosion of the bottom row of tubes in a tube bundle as a function of the vertical distance between the distributor and the bottom row of tubes. These computed results were obtained from the bubble coalescence/ tube erosion computer model using the values n = 3 and Y = 0.25.

elevation of 0.55 m. Finally, the calculated rates of erosion as a function of tube elevation are shown in Figure 10. The erosion rates shown in this figure, which are only for the bottom row of tubes in the bundle, indicate how the erosion rate varies as the distance between the distributor and the bottom row of tubes is changed. At positions close to the distributor, single bubble collisions dominate, with the major contribution to erosion due to the high frequency of single bubbles. The rate of erosion drops at 0.4 m, due to a local decrease in double bubble frequency, and then increases abruptly at 0.5 m as the double bubble frequency rises. Further from the distributor, the rate of erosion decreases again as the bubbles grow larger, the single and double bubble frequencies both decrease and overall rates of collision of bubbles with the tube decrease.

SUMMARY

The sample results from the computer model, presented for the TVA 20 MW pilot plant illustrate the sensitivity of the rate of erosion to the vertical distance between the bottom row of tubes in the tube bundle and the distributor. For this case, the calculated erosion rate is relatively large within 0.35 m of the distributor due to the very high single bubble frequencies in this region. Locally high rates of coalescence between 0.45 and 0.65 m lead to locally high rates of wastage at these locations. At distances further from the distributor, the calculated rate of erosion decreases substantially due to lower bubble frequencies at the higher elevations.

These calculated results suggest that small changes in tube elevation can lead to relatively large changes in erosion rate. They also suggest that the design of the TVA 200 MW FBC inadvertently has provided for the bottom row of tubes to be located within a region of relatively high erosion rates.

Additional work is now in progress to use the model to determine the effects of superficial gas velocity, nozzle spacing and minimum fluidization velocity on erosion rates. The results of these calculations will be described in a follow-on paper.

ACKNOWLEDGEMENTS

This work is funded by the Electric Power Research Institute under RP8006-13.

The authors are particularly grateful to John Maulbetsch and Jeffrey Stallings of EPRI for their support of this activity.

NOMENCLATURE

a	= coefficient in erosion equation
A_{Bed}	= cross-sectional area of bed
D	= bubble diameter
D_B	= average bubble velocity
D_o	= initial bubble diameter
E	= amount of material removed from tube during one bubble/tube interaction
f	= bubble frequency
g	= acceleration of gravity
n	= exponent in erosion equation
N	= number of bubbles
N_d	= number of orifices in distributor
P_{21}	= horizontal component of particulate phase velocity at nose of trailing bubble due to motion of leading bubble
q_{21}	= vertical component of particulate phase velocity at nose of trailing bubble due to motion of leading bubble
q_B	= volume of single bubble
Q_B	= visible bubble flow rate
R	= bubble radius
U_A	= isolated rise velocity
U_B	= absolute vertical velocity
U_{mf}	= minimum fluidization velocity
U_*	= gas velocity
V	= velocity of single bubble
V_{B1}	= horizontal velocity
x_d	= vertical distance between bubble centers
x_i	= vertical position of bubble i
Y	= $Q_B/(U_o-U_{mf}) A_{Bed}$
y_d	= horizontal distance between bubble centers
y_i	= horizontal position of bubble i
Δ	= time interval
Δt_c	= interval of time required for coalescence

Subscripts

o	= superficial gas velocity
1	= leading bubble
2	= trailing bubble
DB	= wake of trailing bubble during coalescence
SB	= single bubble wake

LITERATURE CITED

1. Levy, E. and F. Bayat, "The Bubble Coalescence Mechanism of Tube Erosion in Fluidized Beds," Fluidization VI, ed. J. Grace, L. Shemilt, and M. Bergougnou, Engineering Foundation (1989).

2. Rathbone, R., et al., "Measurement of Particle Velocities and Associated Stresses on Immersed Surfaces in Fluidized Beds," Fluidization VI, J. Grace, L. Shemilt, and M. Bergougnou (Eds.), Engineering Foundation (1989).

3. Zhu, J., et al., "Erosion Causing Particle Impacts on Tubes in Fluidized Beds," Fluidization VI, J. Grace, L. Shemilt, and M. Bergougnou (Ed.), Engineering Foundation (1989).

4. Harrison, D. and L. S. Leung, "Symposium on the Interaction Between Fluids and Particles," London: Institute of Chemical Engineers, 127 (1962).

5. Rowe, P. N., Chapter 4 in: Fluidization, J. F. Davidson and D. Harrison (Ed.), Academic Press, London (1971).

6. Toei, R. and R. Matsuno, International Symposium on Fluidization, Netherlands University Press, 271, Amsterdam (1967).

7. Lin, S. P., AIChE Journal, 16, 130 (1970).

8. Gabor, J. D., Ind. Eng. Chem. Fund., 8 (1), 84 (1969).

9. Clift, R. and J. R. Grace, Chapter 3 in: Fluidization, J. F. Davidson, R. Clift, and D. Harrison (Ed.), Academic Press, London, (1985).

10. Nguyen, T. H., J. E. Johnson, R. Clift, and J. R. Grace, "Prediction of Bubble Distributions in Freely-Bubbling Three-Dimensional Fluidized Beds," Fluidization Technology, ed. D. L. Keairns, Hemisphere, Washington, D.C., I, 205.

11. Farrokhalee, T. and R. Clift, "Mechanistic Prediction of Bubble Properties in Freely-Bubbling Fluidized Beds," Fluidization, J. R. Grace and J. M. Matsen (Eds.), Plenum, New York, NY, 135 (1980).

12. Bayat, F., "A Mechanism of Tube Erosion in Bubbling Fluidized Beds," PhD Dissertation, Lehigh University (1989).

13. Mori, S. and C. Wen, AIChE J 21, 109-115, (1975).

14. Baeyens, J., Personal Communication, (1988).

15. Vincent, R. Q., et al., "Erosion Experience of the TVA 20 MW AFBC Boiler," Proceedings 1987 International Conference on Fluidized Bed Combustion, J. P. Mustonen (Ed.), ASME (1987).

16. Majumdar, S., et al., "A Review of Solid Particle Erosion of Engineering Materials," Argonne National Laboratory Report ANL/FE-88-1 (1988).

HEAT TRANSFER TO VERTICALLY FLOWING DILUTE AND DENSE PHASE GAS-SOLIDS SUSPENSIONS

F. Bentahar ■ Institut de Chimie Industrielle, U.S.T.H.B., B.P. 31, Bab Ezzouar, Algiers, Algeria
Y. Molodtsof and J.F. Large ■ Université de Technologie de Compiègne, D.T.P.S., B.P. 649, 60200 - Compiègne, France
K. Alia ■ Institut de Chimie Industrielle, U.S.T.H.B., B.P. 31, Bab Ezzouar, Algiers, Algeria

Heat transfer to vertically flowing gas-solids suspensions is experimentally investigated for both dilute and dense phase regimes. The test rig includes a test section (20 mm ID, 4.2 m in height) with an electrically heated wall which supplies a constant heat flux to the suspension. The latter consisted of atmospheric air carrying 0.22 mm sand particles.

The results obtained at constant gas flow-rate are very well described by the equation theoretically derived by Molodtsof and Muzyka assuming a *self-similar profiles regime* for dilute phase flow. This equation fits the results even for dense phase flow data. The transition observed in the trend of variation of the heat transfer coefficient coincides with the hydrodynamic transition from the similar profiles regime to the dense phase flow regime.

INTRODUCTION

The first published studies concerning heat transfer between gas-solids suspensions and pipe walls appeared in the litterature in the late 1950's. This pioneering work had mostly in view the use of graphite suspensions as heat carrier in nuclear reactors. Despite the relatively large number of careful experimental studies undertaken in this connection no general explanation for the variation of heat transfer coefficients as a function of solids loading has been achieved whilst the reported results exhibit seemingly contradictory trends of variation. Recent experimental studies concerned with heat transfer from or to a circulating fluidized bed did not help to resolve these discrepancies.

Molodtsof and Muzyka (1,2) developed recently a rigorous theoretical approach based on the General Probabilistic Multiphase Flow Equations (1,2,3). Applied to wall-to-suspension heat transfer in dilute phase flow called *similar profiles regime* this approach leads to an explicit expression of the heat transfer coefficient as a function of the loading ratio which has been confirmed experimentally (2,4). Moreover, this equation predicts all the trends of variation observed by previous authors (1).

According to its authors, the validity of this equation should, however, be limited to dilute phase flow conditions. The aim of this work was to experimentally investigate heat transfer to dense phase suspensions in order to determine the deviations from the predicted behaviour. Our results show that Molodtsof & Muzyka's equation still holds beyond the transition to dense phase flow but with modified coefficients.

PREVIOUS WORK

Most of the work undertaken on vertically flowing suspensions has been reviewed in great detail by Muzyka (2). As long as convective heat transfer between the suspension and the wall is under consideration at the exclusion of radiative heat transfer, the major concern is to describe and predict the trend of variation of the heat transfer coefficient with solids loading in various pipe geometries and flow conditions. Reported results show that in some cases coefficients increase monotonically as a function of solids loading; in other cases there is an initial decline at low solids loading followed by an increase when the ratio of solids to gas flowrates exceeds the value of about 1. Heat transfer coefficients remaining lower than that of the gas alone have also been reported (5) even at relatively high loading. Moreover these dramatic differences in overall results are observed with only moderate differences in overall operating conditions.

Recently published experimental studies of heat transfer to or from a circulating fluidized bed (6,7,8) for gas velocities varying in a limited range did not help to clarify the matter. In his review paper, Grace (9) summarized the situation as follows: "No existing correlations give consistent agreement with the available data."

Considering the very complex mechanisms involved in the transfer of heat between a surface and a suspension, clarification through a comprehensive mathematical description of the phenomena can be expected from theoretical approach. Indeed, early experimental investigations have been, soon after,

---

Correspondence concerning this paper should be addressed to Y. Molodtsof.

followed by theoretical attemps (10,11). They are based upon more or less rigorous and complete energy equations for gas and solids phases. In order to obtain a workable solution several simplifying assumptions are made, among which, that of the uniform distribution of the particles throughout the flow field. The analysis proposed by Matsumoto et al. (12) differs from the preceeding by the attempt to take into account eddy diffusivity. A recent alternative approach by Michaelides (13) considers the suspension as a variable density variable heat capacity one-phase fluid.

Attemps to model heat transfer from circulating fluidized bed (14,15) are not based on Eulerian energy equations but start from a phenomenological picture of the flow described in terms of clusters. However, although in a simplified form, Basu & Nag (15) tried to take into account a basic feature of the flow structure i.e. the non-uniform distribution of the solids over the cross-section.

The theoretical approach proposed by Molodtsof (1) and Muzyka (2) is based upon the rigorously derived General Probabilistic Equations for Multiphase Flow (3). Molodtsof (1) showed that in fully developed flow conditions *self-similar solids concentration profiles* were possible asymptotic solutions to the General Equations, together with solids velocity profiles independent of mean particle concentration and a deformation of gas velocity profiles proportional to the average solids concentration. This solution holds at constant gas superficial velocity, when solids volumetric concentrations tend toward zero. Assuming analoguous *self-similar profiles* for the radial temperature distributions of either gas and particles when the suspension is heated by a constant heat flux at the wall, Molodtsof (1) and Muzyka (2) derived the following equation predicting the variations of the wall-to-suspension heat transfer coefficient as a function of loading ratio.

$$\frac{h_m}{h_0} = \frac{(1 + M C)^2}{1 + a M C + b (M C)^2} \quad (1)$$

According to this equation, the loading ratio M together with the solids-to-gas heat capacity ratio C determine the ratio of the heat transfer coefficient of the suspension h_m to that of the gas alone (h_0) flowing in the same pipe at the same flowrate. The dimensionless parameters a and b in Eqn. (1) are compound shape factors involving dimensionless radial concentration, velocity and temperature profiles. They are generally unknowns as the profiles are unkown. Apart from design parameters (pipe diameter, particle size distribution and physical properties etc.) they only depend upon the superficial gas velocity.

Flow structure measurements by Azzi (16,17) for 60 μm cracking catalyst particles carried by atmospheric air evidenced the existence of the *similar profiles regime* as well as that of a transition toward a denser flow regime (where particle mass flux profiles are no longer self-similar with increasing loading). Besides, Eqn. (1) is in excellent agreement with the heat transfer results obtained by Muzyka (2,4) as well as with the data reported by Tien & Quan (18) for constant wall heat flux conditions (2). Moreover, Eqn. (1) predicts all the above mentioned trends of variation reported by previous experimental investigators (1).

It was, therefore, interesting to investigate experimentally the wall-to-suspension heat transfer characteristics for high loading ratios far beyond the hydrodynamic transition to the dense flow regime in order to determine the range of validity of Eqn. (1) as well as the effect of the transition in the flow structure on heat transfer.

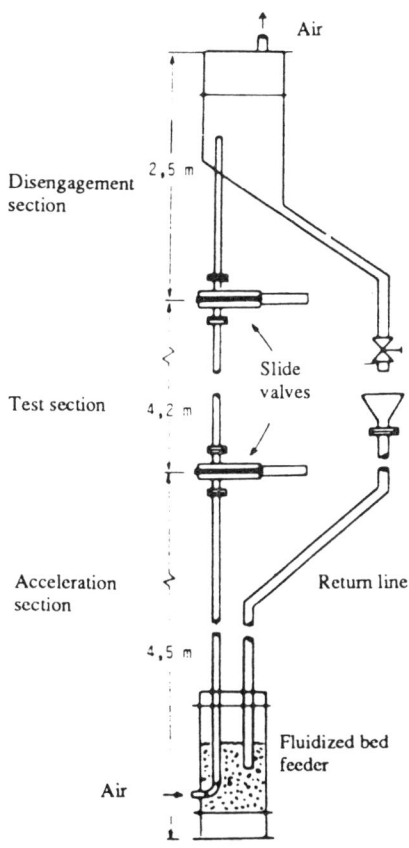

Figure 1. Experimental installation.

EXPERIMENTAL INSTALLATION AND PROCEDURE

The vertical pneumatic conveying system used in this study is shown schematically in Figure 1. The transport line consisted of a 20 mm ID stainless steel pipe approximately 11 m in total height divided into 3 sections : A 4.5 m acceleration zone, a 4.2 m heating section and a 2.5 m disengaging area. Solids were introduced into the flowing gas stream from a fluidized bed feeder 320 mm in diameter and 500 mm in height. The base of the transport line entered the bed and was submerged in the fluidized solids to a height of about 200 mm. Small holes drilled around the circumference of the submerged section allowed the solids to flow from the bed into the line at a rate finely controlled by varying either the pressure above the bed or the bed height. Solids flowrate was determined by collecting the solids leaving the disengaging section over measured time intervals. The gas flowrate was measured by a rotameter placed after the disengaging section. Two pneumatically activated slide valves were placed at the top and bottom of the heating section. The simultaneous closing of these valves allowed the mean solids concentration to be measured.

The heating section consisted of a thin-walled (0.5 mm), 20 mm ID stainless steel pipe specially drawn to ensure even wall thickness. Copper flanges were silver-soldered at each end of the pipe and the pipe was heated by applying a DC supply between the two flanges. Heat input into the system was determined by measuring the voltage across the pipe and the current in the circuit. Wall surface temperatures were measured using platinum resistance thermometers placed at intervals of 250 mm along the pipe length. The heat transfer section was insulated with fiber-glass insulation to minimize the radial heat losses. These losses were determined by calibration as a function of pipe surface temperature. Axial conduction in the pipe was negligible. As overall, losses were less than 10 % of the heat input, the system gave an essentially uniform heat flux to the suspension.

The experiments consisted of starting the gas flow to the transport line followed by the flow of the fluidizing gas to the solids feeder. The flowrates of gas and solids were then set to the desired values, the heat input into the system was fixed and the system was allowed to stabilize over a period of about 2 hours during which flowrates, temperatures and pressures were monitored and recorded. At the end of the experiment, the slide valves were closed and the solids trapped in the test section were recovered and weighed. Sand having a mean diameter of 220 µm and a density of 2620 kg/m^3 was used in all experiments presented here.

The suspension mixed mean temperature profile $T_m(z)$ was calculated via a heat balance. The validity of these computations based on power input, flowrates, suspension inlet temperature and previously calibrated heat losses were checked by the comparison of the calculated outlet mixed mean temperature with the "equilibrium" temperature. The latter was measured by a bare thermocouple protruding within the flow, downstream from the test section, approximately at the middle of the upper part of the pipe which was also thermally insulated. Differences never exceed 1°C in the experiments reported here.

The acceleration section was equipped with 11 pressure taps. The longitudinal pressure profile was recorded during each experiment in order to make sure that the suspension entering the heated section was in fully developed flow conditions. The wall temperature profile given by platinum resistance thermometers was used as thermally fully developed flow criterion : In all experiments, the profiles were linear beyond the first third of the test section, with a slope equal to that of the calculated suspension mixed mean temperature profile. The uniform difference between these two temperature profiles was used as the ΔT dividing the wall uniform heat flux φ to define the wall-to-suspension heat transfer coefficient h_m.

EXPERIMENTAL RESULTS AND DISCUSSION

All the experiments reported here were performed with a superficial gas velocity maintained constant at 10.4 m/s, and varying the solids rate from 7 g/s to 137 g/s. As a result, the loading ratio varied from 1.5 to 34, and, average solids volumetric concentrations ranged between 0.1 % and 2.5%.

Figure 2 shows a plot of the ratio of the heat transfer coefficient of the suspension to that of the gas alone flowing in the same line with the same superficial velocity of 10.4 m/s, as a function of the loading ratio M. One of the typical trends of variation reported by previous authors is observed, with an initial decrease, a minimum heat transfer coefficient recorded for a laoding ratio of about 4.3, followed by a regular increase beyond this minimum.

In order to easily compare our results with the trend of variation predicted by Eqn. (1), a new dimensionless variable Y is defined as follows :

$$Y = \frac{1}{MC}\left(\frac{(1+MC)^2}{(h_m/h_0)} - 1\right) \qquad (2)$$

Using Eqn.(1), the following expression for Y as a function of the product MC can be obtained, which is an alternative form of the equation derived by Molodtsof & Muzyka :

$$Y = a + b\,MC \qquad (3)$$

Figure 3 shows the experimental results obtained in this work in a Y vs. M plot. Since the solid-to-gas heat capacity ratio C was essentially constant in our experiments, an excellent agreement exists, as can be seen, between the results and Eqn. (3), at low solids loading, as previously reported by Muzyka (2). A sharp change in this linear trend of variation is observed about M = 16, but beyond this value the trend is linear again; in other words, Eqn. (3) is also in excellent agreement with our experimental results for M > 16 provided that the constant coefficients a and b for M < 16 are changed in new constants a' and b'.

The theory leading to Eqns. (1) and (3) predicts that the compound shape factors a and b should be constant at low loading for a constant gas velocity, as observed here. Increasing deviations from the predicted behaviour were expected with increasing solids concentration, since the theory is based on the similar profiles regime assumption which fits with the General Equations as average particle volumetric concentration (α_s) tends toward zero. Deviations were expected to result from changes in either flow structure or temperature fields, with respect to their self-similar behaviour.

It is noteworthy that deviations from the trend of variation in the similar profiles regime, occur as a *sharp transition* at a well-defined concentration rather than progressively. An analogous situation has been reported by Azzi and co-workers (16,17) for the flow structure.

When analyzed in a plot of overall variables (i.e. α_s against M) the hydrodynamic transition from the similar profiles regime to dense phase flow regime appears as a *sharp change* in the linear trend of variation observed for a constant gas velocity. Mok et al. (19) reported quite similar behaviours for sand suspensions.

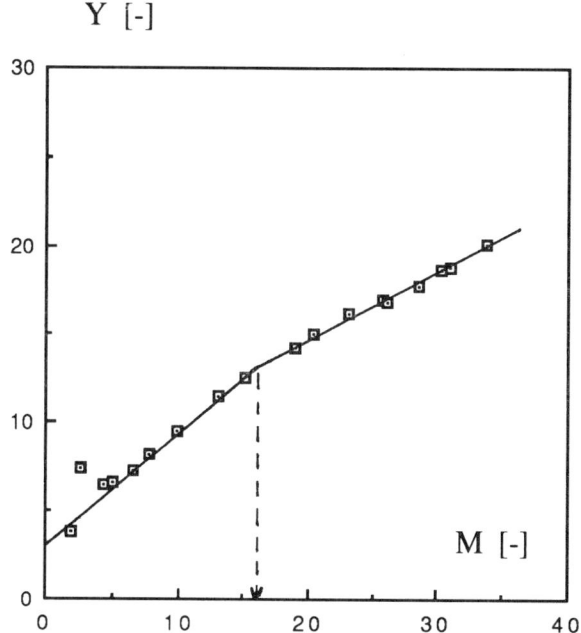

Figure 3. Variation of Y (Eqn. 2) with the loading ratio M at a constant gas velocity of 10.4 m/s

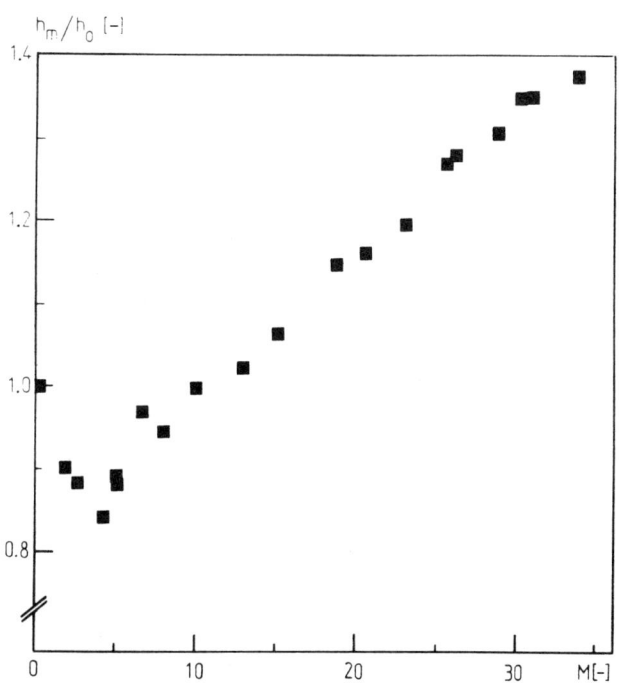

Figure 2. Variation of the heat transfer coefficient ratio with solids loading at a constant gas velocity of 10.4 m/s

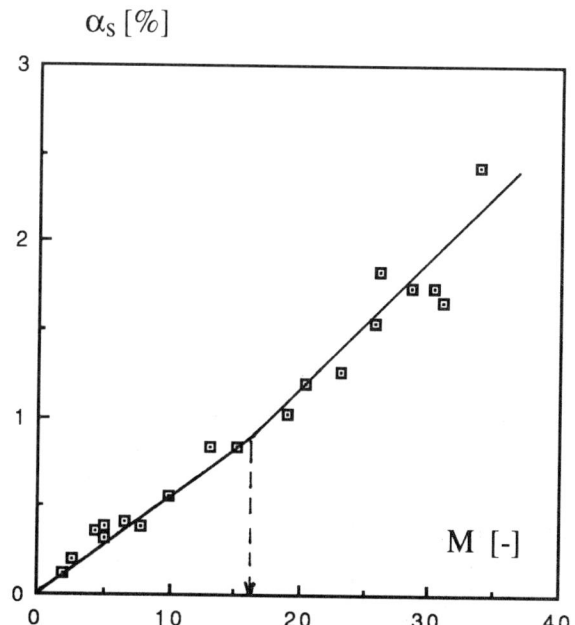

Figure 4. Variation of the solids volumetric concentration with M at a constant gas velocity of 10.4 m/s

The same holds true for our overall hydrodynamic results. Figure 4 shows a plot of the variations of the average solids volumetric concentration measured using the slide valves, as a function of the loading ratio. It can be seen that a sharp transition from the similar profiles regime to dense phase flow occurs at a value of the loading ratio which is *essentially the same* as that of Figure 3. If this coincidence can be generalized to other gas velocities, it could be concluded that the change in the trend of variation of the heat transfer coefficient is due to the regime transition in the flow structure.

The second remarkable feature in Figure 3 is that beyond the transition, the same linear trend of variation holds with different but again constant parameters a' and b'. This fact suggests that some of the noteworthy properties of the profiles still remain beyond the transition, in the dense phase flow regime. Indeed, Azzi (16) observed that, beyond the transition, at a constant gas velocity, the particle mass flux at any radial position varies linearly with the average concentration, with a new but essentially constant slope. This observation parallels the trend of variation of Y as a function of the loading ratio. However, as the dimensionless profiles of concentration, velocities and temperatures are mixed up in coefficients a and b it is difficult to conclude about remaining or new properties of the profiles without a careful analysis by the theory.

This features and coincidence pointed out here are rather promising for the understanding of the hydrodynamic and heat transfer characteristics of gas-solids suspensions. Their generality should, however, be established at least in a significant range of gas velocities prior to an extended interpretation with the help of a theoretical approach.

CONCLUSION

Heat tranfer to vertically flowing gas-solids suspensions has been experimentally investigated in both hydrodynamically and thermally fully developed flow conditions, in dilute and dense phase regimes, for a superficial gas velocity of 10.4 m/s. The results are fairly well described by the Molodtsof & Muzyka heat transfer equation.

The change in the heat transfer behaviour coincides with the hydrodynamic transition from the similar profiles regime toward dense phase flow. According to our results, Molodtsof & Muzyka equation remains still formally valid beyond this transition, but with modified constants. This extended validity suggests that concentration, velocity and temperature fields keep some of their remarkable self-similarity properties in dense phase flow regime.

NOMENCLATURE

a, b   shape factors in Eqn. (1) [-]
C   solids-to-gas heat capacity ratio [-]
h_0 heat transfer coefficient to the gas alone [W/m^2/K]
h_m wall-to-suspension heat transfer coefficient [W/m^2/K]
M   loading ratio : solids-to-gas flowrates ratio [-]
T_m mixed mean temperature of the suspension [K]
Y   dimensionless heat transfer variable defined in Eqn.(2) [-]
α_s solids volumetric concentration averaged over a cross- section [-]
φ uniform heat flux supplied to the suspension [W/m^2]
ΔT reference temperature difference for the definition of h_m [K]

REFERENCES

1. Molodtsof, Y., "Equations Générales Probabilistes des Ecoulements Polyphasiques et Applications aux Mélanges Gas-Solides", Thèse de Doctorat d'Etat, Université de Technologie de Compiègne, Compiègne, France (1985).

2. Muzyka, D. W., "The Use of Probabilistic Multiphase Flow Equations in the Study of the Hydrodynamics and Heat Transfer in Gas-Solids Suspensions", Ph. D. Thesis, University of Western Ontario, London, Canada (1985).

3. Molodtsof, Y. & Muzyka, D. W., Int. J. Eng. Fluid Mech., 2 (1), 1-24 (1989)

4. Molodtsof, Y., Muzyka, D. W., Large, J. F., & Bergougnou, M. A., "The Use of Asymptotic Similar Solutions to Probabilistic Multiphase Flow Equations to Predict Heat Transfer Rates to Dilute Gas-Solids Suspensions", Proc. XVI th. I.C.H.M.T. Symp., Dubrovnik, Yugoslavia (3-7 Sept. 1984)

5. Boothroyd, R. G. & Haque, H., J. Mech. Eng. Sci., 12 (3), 191-200 (1970)

6. Basu, P., Nag, P. K., Chen, B. H. & Shao, M., Chem. Eng. Comm., 61, 227-237 (1987)

7. Wu, R. L., Lim, C. J., Chaouki, J. & Grace, J. R., AIChE J., 33 (11), 1888-1893 (1987)

8. Furchi, J. C. L., Goldstein Jr., L., Lombardi, G. & Mohseni, M., "Experimental Local Heat Transfer in a Circulating Fluidized Bed", in *Circulating Fluidized Bed Technology-II* (P; Basu & J.F. Large Eds.), Pergamon Press, 263-270 (1988)

9. Grace, J. R., "Heat Transfer in Circulating Fluidized Beds", in *Circulating Fluidized Bed Technology* (P. Basu Ed.), Pergamon Press, 63-81 (1986)

10. Tien C. L., *J. Heat Transfer*, 5, 183-188 (1961)

11. Depew, C. A. & Farbar, L., *J. Heat Transfer*, 85, 164-169 (1963)

12. Matsumoto, S., Takahashi, A., Suzuki, M. & Maeda, S., *J. Chem. Eng. Japan*, 12 (3), 183-89 (1979)

13. Michaelides, E. E., *Int. J. Heat Mass Transfer*, 29 (2) 265-73 (1986)

14. Subbarao, D. & Basu, P., *Int. J. Heat Mass Transfer*, 29 (3), 487-89 (1986)

15. Basu, P. & Nag, P. K., *Int. J. Heat Mass Transfer*, 30 (11), 2399-2409 (1987)

16. Azzi, M., "Etude des profils de flux de particules dans l'écoulement vertical établi d'une suspension gaz-solide", Thèse de Doctorat, Université de Technologie de Compiègne, Compiègne, France (1986)

17. Monceaux, L., Azzi, M., Molodtsof, Y. & Large, J. F., "Particle Mass Flux Profiles and Flow Regime Characterization in a Pilot-Scale Fast Fluidized Bed Unit", in *Fluidization V* (K. Ostergaard & A. Sorensen Eds.) Engineering Foundation N.Y., 337-44 (1986)

18. Tien C. L. & Quan, V., *Trans. A. S. M. E.*, 62-HT-15, 1-9 (1962)

19. Mok, S. L. K., Molodtsof, Y., Large, J. F. & Bergougnou, M. A., *Can. J. Chem. Eng.*, 67, 10-16 (1989)

DESIGN OF AN AGGLOMERATION RESISTANT GAS DISTRIBUTOR

R.H. (Hank) Birk ■ The Dow Chemical Co., Central Research, Engineering Research Laboratory, Building 1776, Midland, MI 48674

G.A. (Gary) Camp ■ Dow Chemical USA., Louisiana Division, Plastics Projects Department., Building 807, Plaquemine, LA 70765

L.B. (Loyd) Hutchinson ■ Dow Chemical USA., Louisiana Division, Applied Science and Technology Laboratories, Building 2506, Plaquemine, LA 70765

This paper describes how severe agglomeration problems were eliminated in a newly developing process for reactive modification of a polymer in an industrial fluidized bed. These problems were solved by using a newly derived method to design a gas distributor with horizontal (instead of vertical) jets and inclined (rather than flat) surfaces. Basic principles and fundamental experiments with a non-reacting full-scale transparent model of the fluidized bed reactor were used to develop the new design method. An actual market development plant was operated to experimentally verify the new gas distributor design under normal reacting conditions.

Process Development. The work in this study was part of a larger team effort to develop a commercial batch fluidized process for modification of a specific polymer by reaction of polymer powder with certain gaseous reactants, as shown in Figure 1 below:

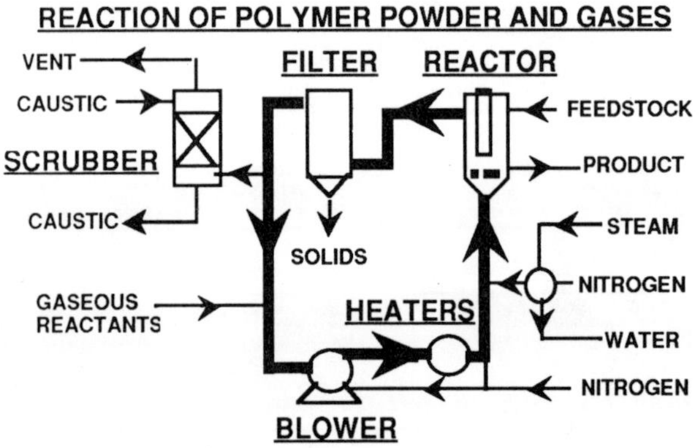

REACTION OF POLYMER POWDER AND GASES

Figure 1. Process Flowsheet of the Market Development Plant for Modification of a Polymer in a Fluidized Bed Reactor.

In this particular process, a batch of the polymer powder was fluidized with a heated recirculating mixture of gaseous reactants diluted by nitrogen with a constant effluent purge to a scrubber. Even at temperatures below the softening point of the polymer powder, the chemical reaction occurred rapidly when free-radical initiators, such as u.v. (ultraviolet) light, were used. Alternative processes typically used homogeneous catalyst(s) in a solution in which the polymer was dissolved. A major economical advantage of the fluidized (solvent-free) process was the lower operating cost that resulted from eliminating the need for solvent removal. However, for the process to be feasible, it was necessary to solve very significant mass and heat transfer problems as follows.

Solids Circulation. Since the u.v. (ultraviolet) light did not penetrate very far into the fluidized bed of powder, there was only a very small portion of the reactor volume near the u.v. light sources that was catalytically active. So in order to overcome this type of mass transfer limitation, it was necessary to design the reactor to obtain excellent circulation of the reactants to and from the u.v. light source zones. Both the particles of solid polymer and the fluidizing mixture of gases were reactants. Without adequate circulation, the production rate would be uneconomical because the desired chemical reaction occurred relatively slowly after the reactants left the small localized u.v. light zone.

Agglomeration. However, outside of the catalytically active zone, even the relatively slow uncatalyzed rate of reaction generated sufficient heat to cause sintering and agglomeration of the polymer if the particles were not fluidized and circulated to dissipate and remove the energy of reaction. This was the primary heat transfer problem and would lead to the formation of large clumps of polymers which were not only a quality control problem but eventually caused premature plant shutdowns due to blockages. These blockages occurred most often on the gas distributor plate.

DESIGN MODEL

Design Concepts

Conventional Designs. In Figure 2 below, the gas distributor was a wire mesh screen used during the early development stage of this process. The pressure drop across the wire screen gas distributor at a constant (and typical) flowrate was plotted as a function of the cumulative reaction time. It was clear from this data that after about 6 hours the pressure drop increased at an approximately exponential rate with operating time. This quickly resulted in a plant shutdown due to the high pressure drop which exceeded the recycle blower's capability.

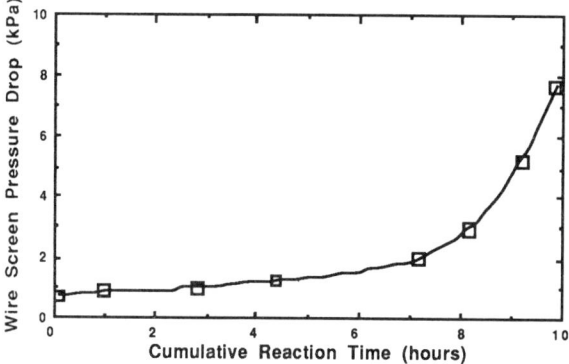

Figure 2. Pressure Drop Across a Conventional Gas Distributor in the Market Development Plant During the Modification of a Polymer in a Fluidized Bed Reactor.

Other types of conventional gas distributors with V.J.F.S. (Vertical Jets and Flat Surfaces), such as plates of porous sintered metal and multiorifice plates, were also used. These other types were also unsuccessful due to severe plugging problems.

Alternative Design. After consulting with several fluidization experts, including Dr. Alan W. Weimer (1) and Dr. Frederick A. Zenz (2), another alternative gas distributor was designed using H.J.I.S. (Horizontal Jets and Inclined Surfaces). Figure 3 pictured a cross-sectional view (not to scale) of the key conceptual features of this alternative gas distributor, which was denoted as the H.J.I.S. design. Notice that the fluidizing gas flowed upward from the lower part (or plenum chamber) through the few larger vertical holes directly under the inclined surfaces (usually referred to as "tents" during the rest of this paper). Then the gas flow was deflected by the "tents" to form jets at the many small horizontal orifices at the base of the "tents". As a result, the flat areas between the base of the "tents" were kept clean by the horizontal jets. This concept has been referred to in the literature, for example by Jenkins, Jones, Jones and Beret (3). However, no published methods for designing a H.J.I.S. distributor were found, so this study was conducted.

Design Equations

The design method was the primary subject of this paper and it was developed from basic principles, fundamental experiments and engineering experiences. The nature of the alternative H.J.I.S. gas distributor, which gave the best performance, was shown in Figure 3.

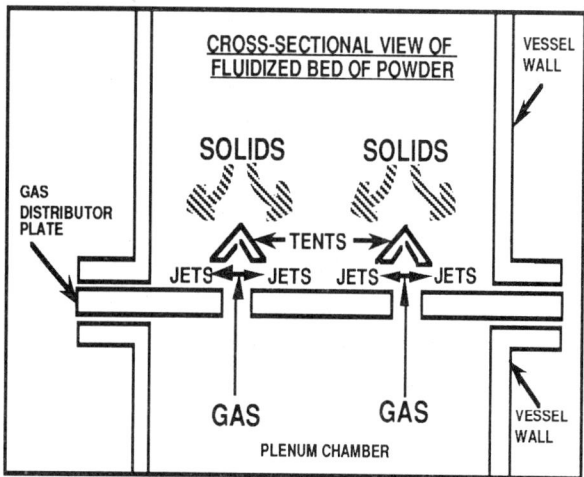

Figure 3. Conceptual Cross-Sectional View of the Alternative Gas Distributor Involving Horizontal Jets and Inclined Surfaces for the Modification of a Polymer in a Fluidized Bed Reactor.

Assumptions. Figure 4 (below) aided in the visualization of the major assumptions for developing the H.J.I.S. distributor design equations. First, all of the horizontal orifices were square and were staggered or offset with respect to the orifices of the adjacent "tents". This had the effect of producing horizontal jets from one tent which essentially fit in between the jets from the adjacent "tent" to provide sweeping coverage of essentially all of the flat surface areas between the "tents". As viewed from below, the jets were triangular in shape and extended from one "tent" to just touching the base of the next adjacent "tent", but did not overlap the surface area of the adjacent "tent". The apex of the triangular jet was visualized as being positioned behind the actual orifice and underneath the "tent". This assumption required that the value of the jet half angle be estimated in order to determine the unseen (or imaginary) distance from the apex of the jet to the orifice. A value of 6.35 degrees was used based on J.M.D. Merry's paper (4) in 1971.

SIDE VIEW

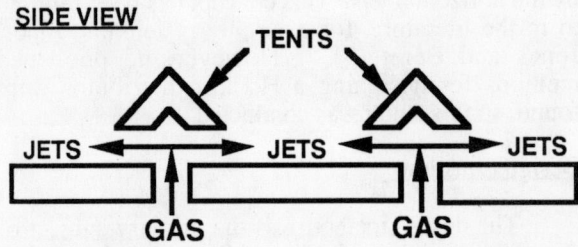

TOP VIEW

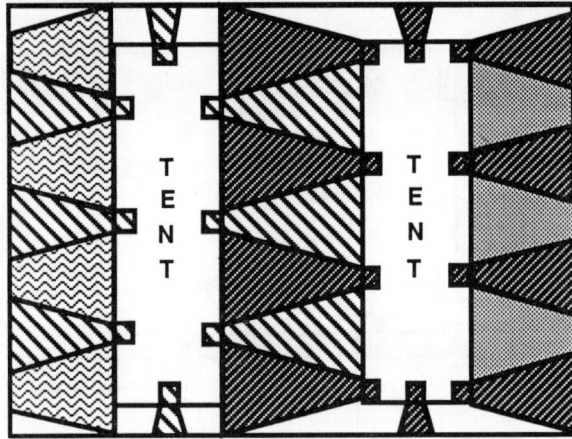

Figure 4. Plan View for Visualization of the Assumptions for Designing the Alternative Gas Distributor Utilizing Horizontal Jets and Inclined Surfaces for the Modification of a Polymer in a Fluidized Bed Reactor.

Also, the height of the "tents" were calculated using a "tent" half angle of about 20 to 30 degrees, which in general depended on the angle of repose for the particular powder being fluidized. These descriptions applied to the orifices and horizontal jets on the longer sides of the "tents". However, on the shorter sides of the "tents", which were adjacent to the cylindrical walls of the reactor, the model neglected the flat surface area covered by the horizontal jets. This was a very conservative assumption which insured that there were no stagnant zones in the flat distributor areas near the reactor walls. But, the number of orifices along both the shorter and longer sides of the "tents" were accounted for in the model to avoid serious errors associated with the effects of the total pressure drop across the gas distributor.

<u>Velocities.</u> Equation (1) for pressure drop of gas through an orifice, as given in standard textbooks such as Kunii & Levenspiel (5), was used in this work to estimate the velocity of gas through the horizontal orifices, as shown below. The value of the gas distributor pressure drop was designed to be equal to about 20 percent of the total pressure drop of the fluidized bed. A value of 0.6 was used for the orifice discharge coefficient because the Reynolds number values were calculated to be greater than 3,000 during this study. Other variables and parameters for all of the equations were defined later in the notation section.

$$v_o = C_D \sqrt{\frac{2 g_c \Delta p_D}{\rho_o}} \quad (1)$$

Equation (2) was used to calculate the normal operating velocity for the gas flowing through the reactor as shown below.

$$v_R = C_F v_{mf} \quad (2)$$

Based on observations during the operation of the fluidized bed, a value of about six for C_F usually insured adequate fluidization without excessive entrainment and carryover of the solid particles. Standard textbooks equations, such as those from Kunii & Levenspiel (5), were used in this work to estimate the value of v_{mf}, which was verified by separate fluidization experiments.

<u>Horizontal Jets.</u> From a mass balance of the total flow of gas through all of the horizontal (square) orifices and through the reactor vessel cross-sectional area, Equation (3) was derived to calculate the total number of horizontal orifices in the gas distributor plate:

$$N_o = \frac{S_R v_R}{S_o v_o} \quad (3)$$

Since the reactor vessel was cylindrical, the shape of the cross-section was circular. The "tents" were positioned above the gas distributor plate and the "tents" varied in length depending on their distance from the center of the gas distributor plate. Relatively larger vertical orifices (approximately 0.006 meters in diameter, spaced about 0.04 meters apart) through the plate directly underneath the "tents" distributed the gas to the smaller horizontal orifices (approximately 0.0007 meters diameter) at the base of the "tents". Gas flowing from the horizontal (square) orifices formed jets which penetrated into the fluidized bed of solid particles.

$$L_J = D_o \left[\left\{ 5.25 \left(\frac{\rho_o v_o^2}{[1-\epsilon_{mf}]\rho_p g \overline{D_p}} \right)^{0.4} \left(\frac{\rho_g \overline{D_p}}{\rho_p D_o} \right)^{0.2} \right\} - 4.50 \right] \quad (4)$$

The length of the jets from the horizontal (square) orifices was calculated using Equation (4) above, which was a correlation by Merry (4). Fluidization experiments verified that Equation (4) was applicable to the particular gases and solid particles used during this work.

Equations (5) and (6) were derived to calculate the width of the horizontal jets. The first parameter was the distance underneath the "tents" from the orifice to the imaginary apex (unobservable source) of the jet, which was calculated by approximating the jet as a right triangle, as shown below:

$$x_o = \frac{L_o}{2 \tan \alpha_J} \quad (5)$$

$$B_J = 2 \tan \alpha_J [L_J + x_o] \quad (6)$$

Based on the work of Merry (4), a value of 6.35 degrees was used for the value of the jet half-angle in Equations (5) and (6), where the half-angle was at the imaginary (unseen) apex of the jet directly behind the horizontal orifice underneath the "tent".

"Tent" Width. To derive Equation (7) below for determining the "tent" width, several expressions were written. First, an expression was written in terms of the jet dimensions to describe the amount of surface area on the flat portion of the gas distributor plate that was covered by the jets. This included a correction to exclude the area of the unseen part of the jet underneath the "tents" from the total jet coverage area.

A second expression was written in terms of the tent characteristics to describe the amount of surface area on the flat part of the plate that did not need to be swept by the jets to prevent agglomeration because that area was covered by the "tents". To reduce computing time, this expression was simplified by not correcting for the number of horizontal orifices on the shorter sides of the "tents". Thus, Equation (7) below provided a first estimate of the value to the width of the "tents" which was later revised by the more accurate Equation (10).

The combined area represented by these two expressions was equal to the total cross-sectional area of the flat portion of the plate, which was written in terms of the vessel diameter.

$$B_{T^*} = \frac{\frac{\pi D_R^2}{2 N_o} - B_J (L_J + x_o) + L_o x_o}{B_J + L_o} \quad (7)$$

Equation (7) was the result of rearranging the combined expressions for the above mentioned areas and solving for the "tent" width, as shown above.

Number of "Tents". In order to avoid polymer stagnation and agglomeration on the flat portions of the gas distributor between the "tents", it was essential to design the gas distributor so that the separation distance between the "tents" was about equal to (or less than) the effective length of the jets. This insured complete coverage (or sweeping) of the flat areas by the action of the jets, which were assumed to be approximately triangular in shape. This assumption was later shown to be satisfactory.

Consequently, when the vessel diameter was visualized as a line perpendicular to the length (not the width) of the "tents", an expression was written for the vessel diameter in terms of the "tent" characteristics and the jet length. Then the expression was solved for the total number of "tents" which were needed for agglomeration resistant operation as shown below:

$$N_{T^*} = \frac{D_R}{B_{T^*} + L_J} \quad (8)$$

Since the value of the "tent" width was only an approximate value due to the use of a simpler model, the value of the number of "tents" from Equation (8) was used as a first estimate in the iterative calculation procedure which was revised later using a more rigorous model with Equations (10) and (11).

Orifices on Shorter Sides of "Tents". One of the first steps in the iterative calculations was the solution of the equation for determining, N_o', the total number of orifices on the shorter sides of all of the "tents". A visualization of the key concepts used to derive Equation (9) [and also Equation (12) later] for calculating the value of N_o' was depicted in Figure 5. It should be noted for Figure 5 that the shorter sides of each "tent" were located near the wall of the reactor vessel. The equation was derived from geometric considerations, as follows. The total length of the shorter sides of all the "tents" was proportional to the product of the total number of "tents", N_T, and the width of the "tents", B_T. One of the simplifying design constraints was that all of the "tents" had the same width, but the "tents" varied in length (not accurately depicted in Figure 5), depending on their position from the center.

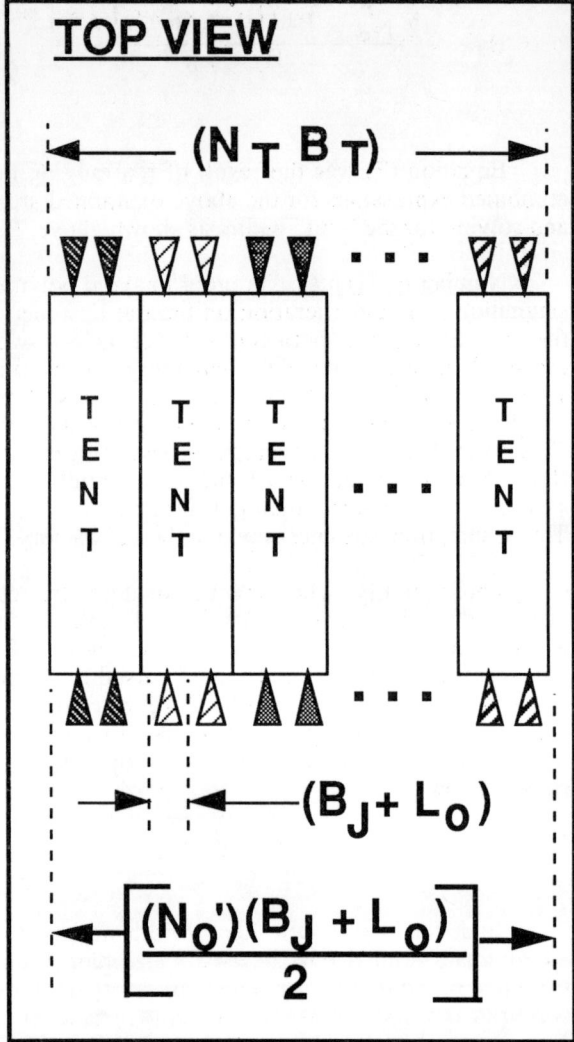

Figure 5. Plan View Showing Concepts Used for Derivation of Equations (9) and (12) for Calculation of the Total Number of Orifices on the Shorter Sides of the "Tents" Near the Reactor Vessel Wall (Not to Scale).

Since each "tent" had two shorter equal sides, the total length of the shorter sides of all of the "tents" was given by $2 N_T B_T$. This expression, $2 N_T B_T$, was also described in terms of the horizontal jets, where B_J was the maximum width of the jet at the end (tip) of the jet, and L_O was both the size of the jet orifice and the separation distance between the jets at their tips. Thus, the distance between each jet orifice was the sum of B_T and L_O, so that the total length of the shorter sides of all of the "tents" was $(N_O')(B_T + L_O)$. The first expression for the total length of the shorter sides of all of the "tents" in terms of the tent characteristics was equated with the second expression for total length in terms of the jet parameters and then rearranged to give N_O' explicitly.

An asterisk subscript was used to denote values that were first estimates based on the simpler model (which did not correctly account for the orifices on the shorter sides of the "tents"), as shown below:

$$N_{o*}' = \frac{2 N_{T*} B_{T*}}{B_J + L_O} \qquad (9)$$

<u>Tent Width (Revised).</u> For revising the estimate of the "tent" width, Equations (10a) and (10b) were derived similarly to Equation (7) above, except that the effect of the horizontal orifices on the shorter sides of the "tents" near the vessel walls were more rigorously accounted for by replacing the parameter, N_O, with $(N_O - N_{o*}')$ or $(N_O - N_O')$ as shown below:

$$B_T = \frac{\frac{\pi D_R^2}{2(N_O - N_{o*}')} - B_J (L_J + x_O) + L_O x_O}{B_J + L_O} \qquad (10a)$$

A value for N_{o*}' was obtained from Equation (9) and then used with Equation (10a) to revise the "tent" width value for iterative calculations with Equations (10b), (11) and (12), as given below. Note that the asterisk subscript was not used with values based on the more rigorous model (which correctly accounted for the orifices on the shorter sides of the "tents").

$$B_T = \frac{\frac{\pi D_R^2}{2(N_O - N_O')} - B_J (L_J + x_O) + L_O x_O}{B_J + L_O} \qquad (10b)$$

<u>Number of Tents (Revised).</u> Equation (11) was identical to Equation (8), except that the accuracy of the value for the total number of "tents" was improved by using the revised values of the "tent" width from Equation (10), which correctly accounted for the effect of the horizontal orifices on the shorter sides of the "tents", as shown below:

$$N_T = \frac{D_R}{B_T + L_J} \quad (11)$$

Number of orifices (revised). Equation (12) was identical to Equation (9), except that revised values of the number and the width of the "tents" from Equations (10b) and (11) were used to improve the accuracy of the number of horizontal orifices on the shorter sides of the "tents", as written below:

$$N_o' = \frac{2 N_T B_T}{B_J + L_o} \quad (12)$$

Iterative Calculations. Next the value of B_T was recalculated using Equation (10b), which was based on the more rigorous jet coverage model, rather than Equation (7), which was derived using the simpler model. Then the value of N_T was recalculated using the revised value of B_T with Equation (11). Next the value of N_o' was recalculated using the revised values of N_T and B_T in Equation (12). This iterative procedure was repeated until the numerical solution converged, i.e. the difference between successive values of N_o' was about 0.1 per cent or less.

"Tent" Locations. Equation (13) was derived to calculate the location of each "tent" on the gas distributor plate in terms of the distance from the center of a specific "tent" to the center of the distributor plate, as given here:

$$x_{Ti} = \psi_{Ti} (L_J + B_T) \quad (13)$$

The values of the function, ψ_{Ti}, depended on two other parameters. The first one was the subscript, i, which denoted the specific "tent", and the second one was the total number of "tents", N_T, which was determined from Equation (11) after the iterative calculations converged to a numerical solution.

For design cases where N_T was an odd number, the first "tent" (i = 1) was located over the center point of the plate and the values of x_{T1} and ψ_{T1} were both zero. In this odd number case, the second (i = 2) and third (i = 3) "tents" were located on either side of the first "tent" (i = 1) with a gap of L_J between the "tents" at their bases and the values of ψ_{T2} and ψ_{T3} were both unity, so that x_{T2} and x_{T3} were both equal to $1(L_J + B_T)$. Likewise, the fourth (i = 4) and the fifth (i = 5) "tents" were located further from the center and on either side of the last two "tents" (i = 2 and i = 3) with separation distances of L_J. Similarly, the values of ψ_{T4} and ψ_{T5} were both two, so that x_{T4} and x_{T5} were both equal to $2(L_J + B_T)$.

In summary, when N_T was an odd number, the values of $\psi_{T1}, \psi_{T2}, \psi_{T3}, \psi_{T4}, \psi_{T5}, \psi_{T6}, \psi_{T7}$ ... were 0, 1, 1, 2, 2, 3, 3 ..., respectively. For computing purposes when N_T was an odd number, ψ_{Ti} was equal to the integer part (decimal portion truncated) of the quantity (i / 2).

On the other hand, for design cases where N_T was an even number, the first (i = 1) and second (i = 2) "tents" were located on either side of the center point of the gas distributor plate. Each of the two "tents" were equally spaced from the center of the plate, with the same gap (L_J) between them as for the odd number cases. Both the values of ψ_{T1} and ψ_{T2} were equal to one-half, which resulted in x_{T1} and x_{T2} having the same value of $(0.5)(L_J + B_T)$. Using the same approach as for the odd number case, the values of $\psi_{T1}, \psi_{T2}, \psi_{T3}, \psi_{T4}, \psi_{T5}, \psi_{T6}$ ... were 0.5, 0.5, 1.5, 1.5, 2.5, 2.5, ..., respectively. For computing purposes when N_T was an even number, ψ_{Ti} was equal to the decimal part (integer portion ignored) of the quantity $[(i + 1) / 2]$ subtracted from the full value of the quantity (i / 2).

"Tent" Length. Equation (14) was derived from geometry. A right triangle was visualized where the hypotenuse was the distance from the center point of the plate to the end of the "tent" near the vessel wall. The length of the hypotenuse was equal to half the value of the reactor diameter, ($D_R / 2$), and the other two adjacent sides of the triangle were the dimensions denoted by ($L_{Ti} / 2$) and x_{Ti}. Based on the Pythagorean theorem, the expression for the length of the hypotenuse in terms of the lengths of the two adjacent sides was rearranged and solved for L_{Ti}, which gave the following:

$$L_{Ti} = \sqrt{D_R^2 - 4 x_{Ti}^2} \quad (14)$$

Orifices on the Shorter Sides of a "Tent". Equation (15), as given below, was based on the assumption that each "tent" had the same number of equally spaced horizontal orifices on the shorter sides of the "tents" near the vessel wall because all of the "tents" were the same width.

$$N_{oi}' = \frac{N_o'}{N_T} \quad (15)$$

Orifices on All the Sides of Each "Tent'". Equation (16) was derived from an expression describing the total length, L_{T_i}, of a given "tent" in terms of the horizontal orifice jet characteristics; B_J, L_o; and the number of horizontal orifice jets along only one of the longer sides of a specific "tent". This number excluded the orifice jets on the shorter sides of the "tent" near the vessel wall and was given by the quantity $[(N_{oi} - N_{oi}') / 2]$. Then N_{oi}' was rewritten in terms of B_T, B_J and L_o, using Equations (12) and (15). Finally the overall expression was solved for N_{oi} as shown below:

$$N_{oi} = \frac{2(L_{T_i} + B_T)}{B_J + L_o} \quad (16)$$

"Tent" Height. Equation (17) was derived from geometry on the basis that all of the "tents" were approximated as triangles with the "tent" height, Z_T, being the hypotenuse of a right triangle and the value, $(B_T /2)$, was the length of the side opposite the half-angle at the "tent" apex, (θ_T).

$$Z_T = \frac{B_T}{2\tan\theta_T} \quad (17)$$

Orifices on the Longer Sides of Each "Tent'". Equation (18) was derived from a balance of the horizontal orifices on any given "tent". The number of orifices on the longer side of a given "tent", which was not necessarily the same for all "tents", was calculated by subtracting the number on the shorter sides from the total number (which included both the shorter and longer sides) as follows:

$$N_{oi}" = (N_{oi} - N_{oi}') \quad (18)$$

FUNDAMENTAL EXPERIMENTS

Equipment

One of the first steps in the fundamental experimental program was to construct and operate a transparent H.J.I.S. gas distributor and fluidized bed vessel. This was used to refine and confirm the theoretically based H.J.I.S. design procedure, which was described above.

Flowsheet. The experimental fluidized bed flowsheet was depicted in Figure 6. The transparent fluidized bed model was built approximately the same size as the reactor in the M.D.P. (market development plant), i.e. 0.61 meters (2 ft.) diameter and 2.74 meters (9 ft.) high. It was designed and used only for operation without u.v. light sources and no chemical reactions were conducted in the transparent model. Actual gaseous reactants were not used in the transparent model of the M.D.P. fluidized bed reactor. Only nitrogen was used to fluidize the powdered reactants. But all the other conditions; such as the gas velocities, the types and quantity of powdered feedstocks; were as close as possible to those in the M.D.P.

Instruments. Also shown in Figure 6, were the pressure regulator, PC1, for reducing the nitrogen pressure from about 1618 kPa (220 psig) to the 377 to 1308 kPa (40 to 175 psig) range; two indicators, FC1 and FC2, for high and low nitrogen flowrates, 0.139 and 0.025 cubic m/s (294 and 53 SCFM) with integral orifice diameters of 1.87 and 0.864 cm (0.735 and 0.340 inches) as well as the indicators for absolute and differential pressure, PI 1 to PI 3 and DPI 1 to DPI 3. Four shutoff valves, V1 to V4, and one flow control valve, V5, were also depicted.

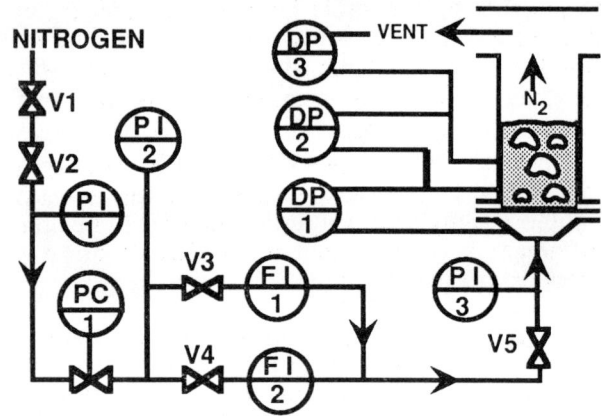

Figure 6. Process Flowsheet of the Transparent Hydrodynamic Model Vessel of the Fluidized Bed Reactor in the Market Development Plant for Polymer Modification Reactions.

Procedures. From underneath the transparent fluidized bed, replicate visual measurements by multiple observers were made of the two-dimensional size of the horizontal jets on the upper surface of the gas distributor plate. Bed height was observed from the side of the vessel. From atop the equipment, the sizes of the bubbles were visually estimated as they erupted at the upper surface of the fluidized bed. Space limitations did not permit the results of all of these measurements to be presented in this paper, but they were documented elsewhere by Birk, Camp and Hutchinson (6). However, one of the first set of fundamental experiments was the determination of the dimensions of the distributor's horizontal jets with a typical powdered feedstock. This information was critical for the correct design of the H.J.I.S. gas distributor.

Data

Horizontal Jets.
In Figure 7 below, the mathematical description of the horizontal jets was simplified to only two dimensions, namely the jet length and the jet width. These two jet dimensions were plotted as functions of the nominal gas velocity through the horizontal orifices. There was no obvious systematic dependence on reactor load size, which varied from 44 kg. (97 lbs.) to 179 kg. (395 lbs.).

Jet Lengths.
J.M.D. Merry (4) correlated the horizontal jet lengths for some materials as a function of the orifice velocity and the physical properties of the powders. Merry's 1971 correlation conservatively described the experimental data (see Figure 7) because the actual measured jet lengths were usually greater than Merry's predictions which would tend to insure adequate coverage or sweeping of the flat surfaces on the gas distributor. This apparent tendency towards underpredicting jet lengths was consistent with observations by others, such as Professor Peter Rowe and his co-workers, that the stability (i.e. the length) of the gas jets can be increased by contact with adjacent surfaces, such as the distributor plate in this study, which were not within the scope of Merry's correlation.

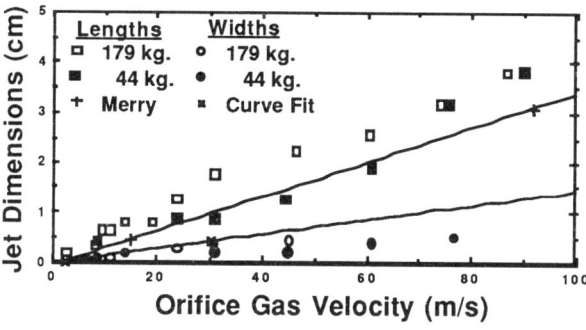

Figure 7. Correlations of the Horizontal Jet Dimensions (Length and Width) as a Function of Orifice Gas Velocity in the Transparent Hydrodynamic Model Vessel of the Fluidized Bed Reactor in the Market Development Plant for Polymer Modification Reactions.

Jet Widths.
For the case of the other dimension, i.e. jet widths, the experimental data (Figure 7) was correlated by linear regression of the jet width in units of centimeters as a function of v_o (orifice velocity in meters/second). The resulting empirical equation was: Jet Width (in centimeters) = $(0.0145) v_o - (0.03353)$. Because of the spacing of the horizontal orifices, the jets could not be wider than about 0.005 m. (0.5 cm.) before impinging on the adjacent jets. This was the reason that the experimental data showed a maximum value of about 0.005 m for the jet width as the orifice velocity increased above values of about 30.5 m./s. (100 ft./sec.).

Therefore, this empirical equation should not be used if the horizontal orifices are spaced more than about 0.005 m apart and also not used for materials which are significantly different from those in this study. It should be noted that this 0.005 m. spacing criterion was satisfied in the final gas distributor design. Also, the experimental values of the jet widths were examined to insure that Equation (6) would be satisfactory for design purposes. In other words, the actual jet widths were about the same or larger than the predicted values from Equation (6).

PLANT SCALE-UP VERIFICATION

Pressure Drop.
In Figure 8, two sets of data were presented. The first set was the pressure drop across a new unused H.J.I.S. gas distributor as a function of superficial gas velocity when there was no powder in the bed, i.e. a clean distributor and an empty reactor (recall from Figure 1). However, the second data set of pressure drop values was taken after that same distributor had been in service for more than 36 hours when the reactor was fully loaded with powder. The success of the new distributor design was obvious because the two sets of pressure drop data were essentially identical. It was concluded that the H.J.I.S. gas distributor design, which had operated at least 36 hours without any significant problems, was the best design that had been evaluated.

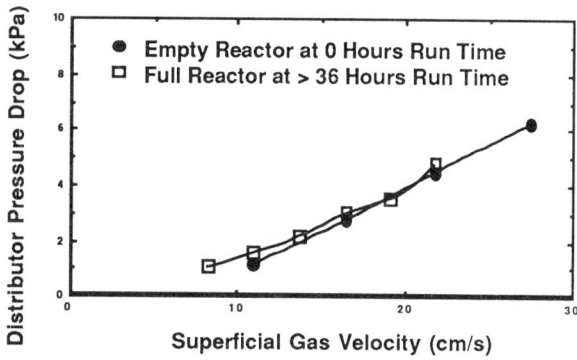

Figure 8. Pressure Drop Across the Alternative (H.J.I.S.) Gas Distributor in the Market Development Plant During the Modification of a Polymer in a Fluidized Bed Reactor.

Comparison.
The H.J.I.S. design was significantly better than the other types; i.e. V.J.F.S.

gas distributors, such as the woven screen (recall from Figure 2); which developed problems serious enough to cause a plant shutdown in less than 10 hours. These difficulties were related to the relatively stagnant regions between the vertical jets, which were inherent to those types of gas distributors. The V.J.F.S. gas distributors were not adequate to prevent agglomeration of the polymer powder without the possibility of uneconomically high fabrication costs and pressure drops associated with several practical limitations.

Although the above data sets illustrated the success of the final H.J.I.S. distributor plate, the first preliminary designs of the alternative type of gas distributor (not shown in this paper) did not perform nearly as well as the final design. However, even the early H.J.I.S. types were so much better than the V.J.F.S. types that the promising potential of the H.J.I.S. type was recognized.

<u>Discussion.</u> With a properly designed H.J.I.S. distributor, the solid polymer particles were not allowed to stagnate on the exposed surfaces of the gas distributor due to two factors. First, the inclined surfaces or "tents" utilized gravity and the angle of repose of the particles to insure that there was no stagnation of polymer above those areas. Second, there was no stagnation in the other flat areas between those "tents" due to the sweeping action of the horizontal jets before they dissipated and their gas entered the bulk of the bed to fluidize the particles. It was found that the use of inclined surfaces permitted a significant reduction in the portion of the surface area of the gas distributor plate which needed to be swept by orifice jets, i.e. shorter jets could be used. One way to view this, was that the capital cost of the fabrication of the "tents" was "paid for" by the reduced operating costs of the shorter jets in the form of lower overall pressure drop across the gas distributor plate.

<u>Conclusions.</u> However, the cost concern was secondary to the chemical reaction performance issue. Reactor performance was very strongly dependent on the rates of mass and heat transfer. A gas distributor design was required which produced sufficient mixing of the gas and the particles to eliminate both the mass and the heat transfer problems. The derived design procedure and equations allowed the size and location of the "tents" and their orifices to be calculated to produce uniformly active jets. This very successfully prevented powder stagnation and gas distributor blockages (recall from Figure 8).

ACKNOWLEDGEMENTS

The authors recognize the importance of the consultations with A.W. Weimer and F.A. Zenz during the early stages of the project, which directed the project towards its successful conclusion. The authors gratefully acknowledge access to the research facilities in the Louisiana Division of Dow Chemical USA and the assistance and support of the following personnel (in alphabetical order): Robert R. Blanchard, Cathy C. Delcambre, Earl George, Ronald L. Glomski, Edward Jones, Irby C. Jones, Duane S. Lehman, Donald E. McLemore, Mark T. Mitchell and Carl J. Stevens.

NOTATION

<u>Roman Letters</u>

B_J = thickness (or width) of jet at its end [m]

B_T = revised estimate of the width of the "tents" (rigorous model) [m]

B_{T*} = first estimate of the thickness (or width) of the "tents" (simple model) [m]

C_D = orifice discharge coefficient [dimensionless]

C_F = ratio of v_R to v_{mf} [dimensionless]

D_o = equivalent circular diameter of the horizontal (square) orifices [m]

$\overline{D_p}$ = average diameter of fluidized particles [m]

D_R = inside diameter of fluidized bed reactor vessel [m]

g = acceleration of gravity [m/s^2]

g_c = gravitational constant [dimensionless]

L_J = length of the horizontal jets (distance from orifice to the end of the jet) [m]

L_o = linear dimension of horizontal (square) orifice [m]

L_{Ti} = length of a specific "tent", as denoted by subscript i, where i = 1, 2, 3, 4 ... N_T [m]

N_o = total number of horizontal (square) orifices [dimensionless]

N_o' = revised estimate of the total number of horizontal (square) orifices on the shorter sides of the "tents" near the vessel walls (rigorous model) [dimensionless]

$N_o^{*'}$ = first estimate of the total number of horizontal (square) orifices on the shorter sides of the "tents" near the vessel walls (simple model) [dimensionless]

N_{oi} = the total number of horizontal (square) orifices on a specific "tent" as denoted by subscript i, where i = 1, 2, 3, 4 ... N_T [dimensionless]

N_{oi}' = the number of horizontal (square) orifices on the shorter sides of "tents" near the vessel wall, as denoted by subscript i, where i = 1, 2, 3, 4 ... N_T [dimensionless]

N_{oi}'' = the total number of horizontal (square) orifices on the longer sides of a specific "tent" as denoted by subscript i, where i = 1, 2, 3, 4 ... N_T [dimensionless]

N_T = revised estimate of the total number of the "tents" (rigorous model) [dimensionless]

N_T^* = first estimate of the total number of "tents" (simple model) [dimensionless]

S_o = cross sectional area of a horizontal (square) orifice [m^2]

S_R = reactor cross sectional area [m^2]

v_{mf} = minimum (or incipient) fluidization gas velocity [m/s]

v_o = superficial orifice gas velocity [m/s]

v_R = superficial reactor operating gas velocity [m/s]

x_o = distance from horizontal (square) orifice to the imaginary apex of the orifice jet (underneath the "tent") [m]

x_{Ti} = distance from the center of the gas distributor to the center of a specific "tent", as denoted by subscript i, where i = 1, 2, 3, 4 ... N_T [m]

Z_T = height of the "tents" (distance from the base to the apex) [m]

Greek Letters

α_J = half-angle of horizontal (triangular) jet [dimensionless]

Δp_D = gas distributor pressure drop [Pascal]

ε_{mf} = voidage of the incipiently fluidized bed [dimensionless]

θ_T = half-angle of the "tent" (at its apex or highest elevation) [dimensionless]

ρ_g = density of gas in fluidized bed [kg/m^3]

ρ_o = orifice gas density [kg/m^3]

ρ_p = density of fluidized particles [kg/m^3]

ψ_{Ti} = a function with values proportional to the value of x_{Ti} (see Equation (13) in text for details) [dimensionless]

LITERATURE CITED

1. Weimer, A.W, (Verbal Communications) Dow Chemical USA, Building 52, Midland, MI, 48667 (1984).

2. Zenz, F.A., (Verbal Communications) Consulting Engineer, P.O. Box 241 - Route 9D, Garrison, NY, 10524 (1984).

3. Jenkins, J.M., III, R.L. Jones, T.M. Jones and S. Beret, U.S. Patent 4,588,790, Assigned to Union Carbide Corporation (1986).

4. Merry, J.M.D., Trans. Instn. Chem.Engrs. (United Kingdom), **49**, 189-195 (1971).

5. Kunii, D. and O. Levenspiel, Fluidization Engineering, John Wiley, New York (1969).

6. Birk, R.H., G.A. Camp and L.B. Hutchinson, Dow Chemical USA, Louisiana Division's Databooks # 3825 and 3895 (1984).

A COMPARATIVE EVALUATION OF NEGATIVELY AND POSITIVELY CHARGED SUBMICRON PARTICLES AS FLOW CONDITIONERS FOR A COHESIVE POWDER

Arunava Dutta and L.V. Dullea ■ GTE Electrical Products, Danvers, MA 01923

The addition of both Aluminum Oxide C (AOC) and Aerosil 200 (A200) in their optimum amounts to a cohesive powder reduces significantly the cohesiveness of the latter. The Hausner Ratio and the quality of fluidization improve remarkably. The optimum amount of either additive is a very small fraction of a monolayer. Charge effects, rather than a spacer or ball bearing mechanism, govern the operation of these flow conditioners. The flow conditioning ability of A200, which has a strong affinity for negative charge, is superior to that of AOC which has an affinity for positive charge. As compared to AOC, the use of A200 results in a lower optimum concentration, a closer approach to ideal bed pressure drop, a significantly lower elutriation loss and a comparable bed expansion.

Fluidization of fine particles, 25 μm or less, is of significant industrial importance. Unfortunately, these materials are the most difficult to fluidize of all powders. Fine particles are characterized by cohesive behavior. Interparticle forces, van der Waals forces and the like, for fine powders are comparable to or greater than the gravitational forces on these particles. Skin drag due to the upward flow of gas through a bed of fine powder is usually insufficient to collapse the structure of the material. In order to obtain an adequate degree of fluidization in beds of cohesive powders, researchers have resorted to different means: addition of small amounts of fluidizing aids to the powder, use of agitators in the fluid bed to break up channels and use of vibrators on the outside of the fluid bed wall.

Dutta et al. ([1]) studied the effects of the addition of highly dispersed alumina (AOC) on the cohesiveness and fluidization of a powder displaying Group C behavior. The fluidizing aid was V blended with the powder in varying amounts, with a

GTE Electrical Products, Danvers, MA 01923

maximum loading of 1.1% by mass of the powder. The authors report that the loosely packed bulk density (LPBD) and the tapped density (TD) of the powder increase with the concentration of AOC up to a certain level of the additive, beyond which no detectable changes are observed. If the compressibility of the mixture were calculated, compressibility being defined as (1 - reciprocal of the Hausner Ratio), it would progressively decrease with AOC addition with a subsequent weak dependence on further additions. The lowest Hausner Ratio, and the minimum cohesiveness, occurs at an AOC loading of 0.35%. Dutta et al. ([1]) also found strong correlation between the Hausner Ratio and fluid bed performance in that, the best fluid bed behavior of the powder was also realized at an AOC level of 0.35%. The bed expansion and solids circulation rate peaked at this concentration of the fluidizing aid.

Steeneken et al. ([2]) discuss the influence of Sipernat 22S, a precipitated spray dried silica, on the mechanical properties of potato starch powder. Addition of about 1000 ppm of the flow aid turns the cohesive starch into a free flowing powder. This is accompanied by a decrease in the porosity and compressibility. Addition of large amounts of flow aids has been reported to sometimes have adverse effects. This is attributed to manifestation of the very low packing density of the conditioner at large loadings. Hollenbach (3) reports a decrease in bulk density of the mixture at levels of conditioner beyond an optimum concentration. Degussa Corporation (4) reports that the effectiveness of Aerosil 200 as a flow agent in the grinding of sulfur exhibits a peak, and the flowability decreases when the percentage of Aerosil exceeds this optimum.

Visser (5) mentions the use of spacers between interacting particles to reduce the van der Waals attractive force, and thereby decrease the cohesiveness of the powder. Spacers could include fine particles or adsorbed gas molecules.

Several highly dispersed materials, which may be useful as flow conditioners, are produced by pyrogenic techniques. These methods of manufacture, flame hydrolysis for example, invariably lead to products which are electrostatically charged. The AOC used by Dutta et al. (1) is reported by Degussa Corporation (6) to have a strong tendency to acquire a positive charge. No fundamental investigations have, however, been reported on the effects of the charge polarity on the efficacy of flow aids. It is the purpose of this paper to make a comparative evaluation between positively and negatively charged submicron particles, as regards their flow conditioning capacity for the same cohesive powder.

THE COHESIVE POWDER

The cohesive powder under investigation is a phosphate based nonporous material with a particle density, determined by pycnometry, of 3100 kg/m^3 and a BET surface area of 530 m^2/kg. This gives a specific surface area based diameter of 3.65 μm. These parameters position this powder in the Geldart (7) Group C classification. In addition, the Hausner Ratio, determined on a Hosokawa Powder Tester, of 1.94 for this material shows that it is highly cohesive.

Particle size distribution, obtained by laser scattering, of the material indicates a volume based mean diameter of 15.94 μm; 90% and 10% by volume are less than 27 and 7.24 μm, respectively. The instrument calculated surface area to volume ratio gives a mean diameter of 12.76 μm. Scanning electron micrographs (SEM) of the powder reveal deviation from a spherical morphology. The particles have two major axes, one of which is much larger than the other.

THE FLOW CONDITIONERS

Two highly dispersed materials from Degussa were chosen for this study: Aluminum Oxide C (AOC) and Aerosil 200 (A200). The former is known, Degussa Corporation (6), to have a strong affinity for positive charge. In addition, the tendency of A200 to develop a negative surface charge has been documented. Using a Keithley electrometer 601, Degussa Corporation (8) measured a charge of about -7.5E-8 C/kg for A200. In addition, a Herfurth Statometer H 1407 was used to measure electrostatic potentials for powder mixtures. Sulfur, for example, with 0.5% A200 was measured at -88V while sulfur with 0.5% AOC had a potential of +28V.

The AOC used has a BET surface area of 84.2E3 m^2/kg, and Degussa Corporation (6) reports an average primary particle size of 20 nm. The particle size distribution is very narrow. The material, a fluffy white powder, is prepared by the flame hydrolysis of anhydrous aluminum chloride. X ray diffraction indicates that AOC is primarily gamma alumina. The particle density of AOC is in the range of 3500 to 3900 kg/m^3. The powder is nonporous. AOC reacts very weakly with most chemicals. At 50% relative humidity, the equilibrium absorption of water by AOC at 293K is about 3% by mass of the powder. The surface of this material contains both Al-OH and Al-O-Al bonds, the latter dominating in number.

The A200 is a high purity silica produced by the hydrolysis of silicon tetrachloride in a hydrogen-oxygen flame. The material used has a BET surface area of 199E3 m^2/kg, and Degussa Corporation (9) reports an average primary particle size of 12 nm. The particle size distribution is very narrow. X ray diffraction reveals an amorphous structure. The density of A200 is about 2200 kg/m^3, and the particles are nonporous. At

50% relative humidity, the equilibrium absorption of water by A200 at 293K is about 2.4% by mass of the powder. The surface of the material has both siloxane and silanol groups, the latter imparting a hydrophilic nature to the surface.

EXPERIMENTAL METHODS AND SCOPE

Blending of the Flow Conditioner and the Cohesive Powder

A Patterson-Kelley twin shell pin intensifier blender, model IB-4, was used to blend the flow conditioner with the cohesive powder. The working volume of the blender is 0.0038 m^3. The following steps were followed in sequence to make up a total of 2100s of mixing time: 600s with the intensifier bar, 600s without the bar, 300s with it, 300s without and a final 300s with the intensifier bar activated.

Twelve concentrations each of AOC and A200 were chosen for this study: 0.003125%, 0.00625%, 0.0125%, 0.025%, 0.05%, 0.1%, 0.2%, 0.35%, 0.55%, 0.8%, 1% and 1.25% by mass of the cohesive powder. The control sample did not have any flow conditioner. Each blended lot was divided into two fractions, one for testing on the Hosokawa apparatus and the balance for fluid bed studies.

Measurements using the Hosokawa Powder Tester

The Hosokawa Powder Tester is currently regarded, Geldart ([10]), as a very useful apparatus for reproducibly measuring the physical characteristics of powders. The incorporation of automation has eliminated much of the human error typically present in such experiment. This instrument was used to measure the loosely packed bulk density (LPBD) and the tapped density (TD) of the samples referred to in the previous paragraph. The Hausner Ratio (HR) is calculated as the ratio of TD to the LPBD. The value of HR always exceeds unity. A drop in the HR may be interpreted, Dutta et al. ([1]), as a decrease in cohesiveness of the powder.

Five runs were conducted on each sample for the LPBD and the TD, over the full range of flow conditioner loading. This leads to five independent estimates for the HR for each sample. The timer switch used during the measurement of TD was set to the full length of 180s.

Statistical analysis was conducted on the LPBD, TD and the HR data. Optimum concentrations for the AOC and A200 were derived from this analysis.

Fluid Bed Experiments

Figure 1 shows the experimental setup used for fluid bed testing of three samples: the control lot with no flow conditioner and the two samples with optimum concentrations of AOC and A200. Fluidizing air was fed through a PFD 301 mass flow controller (MFC) to a 0.1016 m ID and 0.864 m high Pyrex tube which forms the wall of the fluid bed. The MFC has a full scale of 0.833E-3 m^3/s (50 slm) with an accuracy of $\pm$ 1% of span. The set point for the flow was input to a PFD 944 power supply linked to the MFC. The time taken by the MFC for a 10% to 90% change in response to a step in the setpoint is 3s. A 0-5 V output proportional to the air flow rate is available from the PFD 944. The distributor was a 5 micron porosity and 0.00157 m thick SS 304 disc. Attached to the plenum was a Heise model 620 optical pressure transducer, with a range of 0 to 12,291.2 Pa gauge pressure or 0 to 50 inches H_2O gauge pressure. The output of this transducer is 0 to 5 V in this pressure range. The accuracy of the instrument is within $\pm$ 0.15% of span, and the response time is about 2 ms to a step change in pressure.

The MFC and pressure transducer were interfaced using Metrabyte I/O boards to an IBM 7531 computer for real time data acquisition. LabTech Notebook and RS/1 were used as the software.

The fluid bed studies were divided into three parts. The first study focused on the pressure drop versus flow rate characteristics for a sample. After a certain mass of the material had been weighed and placed in the Pyrex tube, a low set point of about 0.7% of full scale on the PFD 944 was used. The set point was increased to about 85% of full scale over at least 10 increments. Each set point was maintained for at least 300s after steady state had been reached.

The frequency of data acquisition was 1 Hz over the full range of flow.

At the end of this period, the second phase of the investigation was initiated. The fluid bed was operated for about 3600s at the highest flow rate established in the earlier experiment. The data collection rate was maintained at 1 Hz. The aim of this part of the study was to find out if significant changes occurred in the fluidization of such materials over long times. The effects of interest here were the bed pressure drop, elutriation losses from the bed, bed particle size distribution and surface coverage of the cohesive powder by the flow conditioner.

Finally, with the flow rate still at the level used in phase II, the data acquisition rate was increased to 20 Hz for a period of 300s. The object of this was to obtain the Fast Fourier Transform (FFT) of the plenum pressure fluctuations, and the power spectrum of the transform. Differences among the samples, if any, in the dominant frequencies can be easily studied by this approach.

RESULTS AND DISCUSSION

Addition of AOC: Effect on LPBD

Table 1 shows the results of the LPBD measurements as a function of the concentration of AOC. Each column reports five independent LPBD values for the same loading of the flow conditioner. Figure 2 is a plot of the mean LPBD of the five runs versus the AOC level.

The Tukey's Post Hoc Test, a statistical test for pairwise comparison between treatments as discussed in Wheeler (11), was conducted on the data presented in Table 1. Confirmatory analysis was used to minimize the occurrence of type I error. The biased pooled variance estimate is recommended as being the correct estimate to use for the standard deviation of the variable. A significance level of 5% was adopted. Any two AOC subgroup averages (there are 13 of them corresponding to the 13 different AOC concentrations) that differ by more than the Tukey's Honestly Significant Difference (HSD) are detectably different at a 5% significance level.

It is worth noting that if treatments X and Y are not detectably different, and so are treatments Y and Z, it does NOT follow automatically that X and Z are not detectably different.

The LPBD increases with AOC addition up to an AOC level of 0.35%, where the LPBD reaches a maximum. This reflects an increase of 33.16% from the mean value for the control sample (no AOC). The LPBD drops slowly when the AOC loading exceeds 0.35%, the decrease being about 5.6% at the highest AOC level of 1.25%. No detectable differences exist between 0% and 0.003125% AOC, between 0.1% and 0.2% AOC, between 0.55% and 0.8% AOC and between 1% and 1.25% AOC.

Addition of AOC: Effect on TD

Table 2 shows the results of the TD measurements as a function of the concentration of AOC. Figure 3 is a plot of the mean TD versus the AOC level.

Though the TD increases with AOC addition up to an AOC concentration of 0.1%, this maximum reflects an increase of only 4.2% from the mean value for the control. The effect of AOC is clearly more pronounced on the LPBD than on the TD. Beyond an AOC level of 0.1%, the TD starts to decrease. The decrease with respect to the maximum is about 5% at the highest AOC loading. Besides experiencing much less of an increase than the LPBD, the TD variation with AOC also shows another interesting feature. No detectable differences exist between more subgroups here than in the LPBD case: between 0%, 0.003125% and 0.00625% AOC, between 0.00625% and 0.0125% AOC, between 0.025% and 0.05%, between 0.05% and 0.1%, between 0.2% and 0.35% AOC and finally between 0.55% and 0.8% AOC.

Addition of AOC: Effect on Hausner Ratio

Table 3 lists the Hausner Ratio (HR) for each of the five runs for all thirteen concentrations of the flow conditioner. The mean HR is plotted versus AOC loading in Figure 4.

The HR and hence the powder cohesiveness decreases with AOC addition up to an AOC level of 0.35%. This is the optimum concentration of

AOC. No detectable differences exist in the HR between samples with 0.1% and 0.2% AOC, between 0.35%, 0.55% and 0.8% AOC and between 1% and 1.25% AOC. The cohesiveness increases when the level of the flow aid is increased beyond 0.8%. The minimum HR of 1.508 corresponds to a 22% reduction with respect to the control sample. The AOC concentration where the HR is a minimum coincides with that at which the LPBD is a maximum. This is in accordance with the results of the previous sections: since the TD does not change much on AOC addition, it is the LPBD peak which decides the minimum in the HR.

Addition of A200: Effect on LPBD

Table 4 lists the results of the LPBD measurements made with the addition of A200. Figure 5 shows the mean LPBD versus concentration of A200.

The LPBD increases with A200 addition (a small drop occurs at 0.00625% A200 loading) up to an Aerosil level of 0.2% where the LPBD reaches a maximum, up 31.87% from the value for the control sample. The LPBD decreases as the A200 loading is increased beyond 0.2%, dropping by about 14.3% (with respect to the maximum) at the highest A200 concentration of 1.25% by mass of the cohesive powder. Statistical data analysis indicates that no two pairs of A200 treatments are the same. This is in contrast to the AOC case where no detectable differences were found between four different pairs.

Addition of A200: Effect on TD

Table 5 presents the TD measurements as a function of the loading of A200. Figure 6 displays the mean TD versus the concentration of the flow conditioner.

The TD experiences an initial decrease when 0.003125% A200 is added. No detectable differences exist between the TD for this treatment and that for the next higher A200 level of 0.00625%. Beyond the latter loading, the TD increases with A200 addition up to an additive level of 0.1%. The maximum TD is only 4.7% above the mean TD value for the control sample. This behavior is similar to that observed with AOC: the effect of the additive is more pronounced on the LPBD than on the TD. No detectable differences exist in the TD between the 0.1% and 0.2% A200 treatments. The TD decreases beyond an A200 level of 0.2%, the decrease being about 9.8% at the highest concentration of the additive.

Addition of A200: Effect on Hausner Ratio

Table 6 shows the five independent values of HR for each concentration of A200 over the full range. Figure 7 is a graph of the mean HR versus A200 loading.

The HR and hence the powder cohesivity decreases (a local peak occurs at an A200 concentration of 0.00625%) with A200 addition up to an A200 level of 0.2%. This is the optimum concentration of A200. The minimum HR at this point is 1.54, down 20.8% from the mean value for the control sample. As in the case of AOC, the concentration where HR is a minimum is also that at which the LPBD peaks. Beyond 0.2% A200, the HR displays an increasing trend. No detectable differences exist between the HR for the following treatments: 0.2% and 0.35% A200, 0.35% and 0.55% A200, 0.55% and 0.8% A200 and finally 1% and 1.25% A200.

A Theoretical Model for the Flow Conditioner Loading

A theoretical estimate of the range of flow conditioner loadings is presented. With reference to Figure 8 the secondary particle diameter, d, is the average size of the additive as it sits on the cohesive powder, and can be estimated from SEM of the mixture. It is important to distinguish it from the smaller primary particle diameter as observed on a TEM of the flow conditioner itself. In this work, d is estimated as ranging between 0.1 and 0.3 μm. In contrast, the primary particle diameters for AOC and A200 are about 20 nm and 12 nm, respectively. AF is lower than the BET surface area of the flow conditioner because d is much larger than the primary particle diameter of the additive. AF is estimated from the product of the ratio of the primary to secondary particle diameters and the BET surface area. The spacing of the flow conditioner particles as they sit on the cohesive powder particles is accounted for by n. The loading, L, is the ratio of mass of flow conditioner to the cohesive powder. The higher the value of n, the smaller is L.

It can be shown that,

$$L \simeq (3.142 \cdot YF \cdot d)/((n+1) \cdot YC \cdot D)$$

It is convenient to express the loading in terms of the relevant specific surface areas,

$$L \simeq (3.142 \cdot AC)/((n+1) \cdot AF)$$

Figure 9 is a plot of L, expressed as a percentage, versus n for both the flow conditioners. In the estimate for AF, d has been taken to be 0.2 μm. The value of n has been varied from a low of 1 to a maximum of 5. The maximum coverage, n=1, of the cohesive powder corresponds to values of L of about 9.88% and 6.97% for AOC and A200, respectively. The experimentally obtained optimum loadings, 0.35% and 0.2% for AOC and A200, respectively, lie well below these maxima. If these optima were plotted on Figure 9, they would correspond to n = 56 and 69 for AOC and A200, respectively. This proves that optimum flow conditioning is achieved at very low surface coverage, much less than a monolayer, of the cohesive powder. At such low coverages the spacer mechanism, Visser (5), of flow conditioners is questionable. In addition a ball bearing mechanism mentioned in Steeneken et al. (2) is doubtful when the substrate coverage by the conditioner particles is so small. Rather, these results indicate that charge effects of AOC and A200 play a very important role in governing their flow conditioning efficacy.

Fluid Bed Studies: Distributor Pressure Drop

The pressure drop across the distributor plate as a function of the air flow rate is shown in Figure 10. All flow rates refer to standard conditions of 294.3K and 0.1 MPa. A flow rate of $0.7E-3$ m^3/s corresponds to a gas superficial velocity of 0.086 m/s. The line of best fit to the data is also presented in Figure 10. This correlation has been used to derive the bed pressure drop for a given total pressure drop. The latter refers to the sum of the distributor and the bed pressure drops.

Fluidization of the Cohesive Powder: No Flow Conditioner

At a flow of about 5% of full scale (FS), the whole powder mass of 1.65 kg travelled up about 0.1 m in the tube as a slug. At 10% of FS, a few layers broke off from the bottom of the slug. Cracks or channels were visible in the bottom bed. At 15% FS, the upper mass collapsed leading to a single bed in the tube. As the flow was increased, more cracks were observed. Movement of the powder was very little. Elutriation of fines started at about 40% of FS. The bed expansion at the highest flow rate was a poor 27%.

Figure 11 displays the relevant pressure drops as a function of the flow rate. The bed pressure drop rises very rapidly as the flow is initiated, then displays a drop before rising again followed by a subsequent gradual decrease as the flow rate is increased. The maximum bed pressure drop, 1753.7 Pa, occurs at about $0.5E-3$ m^3/s and decreases to 1697.6 Pa at about $0.7E-3$ m^3/s. These figures, which are a measure of the quality of fluidization, correspond to 87.8% and 85% of the ideal bed pressure drop. The latter is computed from a knowledge of the mass of powder charged to the fluid bed at the start.

Analysis of the total pressure drop behavior over a period of 3600s at the highest flow rate indicates that the standard deviation is about 1.5% of the mean. The mean and the median differ by about 9 Pa. The mean total pressure drop in the first 300s (4467.2 Pa) exceeds that in the last 300s by about 143.5 Pa (0.58 inch of H_2O). No detectable differences exist in the bed particle size distributions before and after this experiment. The measured loss of material due to elutriation is about 20% of the bed mass before the start of this phase.

Figure 12 shows the power spectrum of the Fast Fourier Transform (FFT) of the plenum pressure fluctuations. The frequencies lie between zero and the Nyquist frequency. The latter is half the data acquisition frequency or 10 Hz. A strong DC component (less than 1 Hz) is visible above the background noise, and there is activity in the 5 to 7 Hz frequency range with a dominant frequency at about 6.5 Hz.

Fluidization with 0.35% AOC

The mass, 2.6 kg, of material charged to the tube rose up in a slug about 0.08 m at a flow rate of 5% of FS. At 10% FS, the slug broke up completely and fell to the bottom of the tube. Bed expansion increased quickly with flow rate up to a flow of about 25% of FS, and then changed only very slowly. Very few channels were observed in the bed all the way up to the maximum flow rate. The bed expansion at the highest flow was about 90%. Beyond about 60% of FS, the surface of the expanded bed looked like a class A bubbling bed.

Figure 13 shows the relevant pressure drops as a function of flow rate, as the latter is increased from zero to about 0.7E-3 m^3/s. As the flow is initiated, the bed pressure drop rapidly undergoes almost a step change followed by a decrease with increasing flow rate. The maximum bed pressure drop, 2845.1 Pa, occurs at about 0.2E-3 m^3/s. The corresponding value at about 0.7E-3 m^3/s is 2618.3 Pa. These correspond to 90.3% and 83.1% of the ideal bed pressure drop.

No detectable differences exist in the mean and median of the total pressure drop as measured over a 3600s period. The standard deviation is 2.24% of the mean. The mean total pressure drop for the first 300s of this period (5351.1 Pa) exceeds that in the last 300s by about 326 Pa (1.32 inch H_2O). SEM indicates no significant difference between the initial and final bed material as regards surface coverage of the cohesive powder particles by the AOC. The elutriation loss is lower than that for fluidization with no additive, amounting to about 17% of the initial bed mass. Comparison of the bed particle size distributions before and after fluidization does not indicate any definite trend. Figure 14 displays the power spectrum of the FFT of the plenum pressure fluctuations. In contrast to the power spectrum for the cohesive powder, the DC components are not strong for this case. Several dominant frequencies, rising above the background noise, are also visible between 4 and 5 Hz and one at about 6.25 Hz.

Fluidization with 0.2% A200

When the flow was initiated at 5% of FS, the 2.2 kg of bed material slugged but quickly broke down to give a single mass at the bottom of the tube. Significant bed expansion was observed at this low flow rate. As the flow rate was increased, the bed height rose though less fast. No appreciable change occurred in the expanded bed height beyond a flow of about 25% of FS. The bed expansion at the highest flow was about 91%. Very few channels could be seen along the tube circumference. The top of the bed was very turbulent. The bed surface behaved almost like a liquid at flow rates exceeding 30% FS.

Figure 15 shows the relevant pressure drops as a function of flow rate. The bed pressure drop displays a trend qualitatively very similar to that obtained in the AOC case. The maximum bed pressure drop, 2600.8 Pa, occurs at 0.2E-3 m^3/s and decreases to 2475.4 Pa at about 0.7E-3 m^3/s. These correspond to 97.5% and 92.8% of the ideal bed pressure drop.

Analysis of total pressure drop data collected over 3600s at the highest flow indicates that the mean and median are not detectably different. The standard deviation is 1.15% of the mean. The mean total pressure drop for the first 300s (5270.2 Pa) exceeds that for the last 300s by a small 67 Pa. SEM indicates no significant change in cohesive powder surface coverage by the A200 between the initial and final bed samples. This supports the finding of Dutta et al. ([1]) that the additive in spite of its much smaller secondary particle size is not elutriated preferentially to the larger cohesive powder. The forces holding the additive to the cohesive powder particles are stronger compared to the drag forces in the fluid bed. Compared to the other two cases, elutriation losses are the lowest while fluidizing with A200: 10.9% of the initial bed mass. This has important industrial consequences, since one desires a fluidizing aid which can provide optimum flow conditioning with the lowest elutriation loss. Table 7 summarizes the elutriation losses for the three cases. No definite trend is observed when comparing the bed particle size distributions before and after fluidization.

Figure 16 is the power spectrum of the FFT of the plenum pressure fluctuations. Dominant frequencies, above the background noise, can be detected at about 3.75, 5.25 and 7.5 Hz. There seems to be fewer clearly dominant frequencies in the 4 to 5 Hz range in this spectrum than in the AOC case.

Table 8 summarizes the effects of the two flow conditioners with the cohesive powder as the control.

CONCLUSIONS

The addition of both AOC and A200 in their optimum amounts to a cohesive powder reduces significantly the cohesiveness of the latter. The Hausner Ratio and the quality of fluidization improve remarkably. In addition, the elutriation rate is reduced. The optimum amount of either additive is a very small fraction of a monolayer. Charge effects, rather than a spacer or ball bearing mechanism, govern the operation of these flow conditioners. Results indicate that the flow conditioning ability of A200, which has a strong affinity for negative charge, is superior to that of AOC which has an affinity for positive charge. The use of A200, which has a lower optimum concentration than AOC, results in a closer approach to ideal bed pressure drop and a significantly lower elutriation rate with a bed expansion comparable to that for AOC.

NOTATION

D cohesive powder specific surface area based mean diameter (μm)
d flow conditioner secondary particle diameter (μm)
n spacing of conditioner particles, multiple of d
AC BET surface area of the cohesive powder (m^2/kg)
AF specific surface area of the flow conditioner (m^{2}kg)
YC particle density of cohesive powder (kg/m^3)
YF particle density of flow conditioner (kg/m^3)
L loading of flow conditioner

LITERATURE CITED

1. Dutta, A. and Dullea, L.V., Paper presented at the AIChE Annual Meeting, Paper No. 163f, Washington, D.C., 1988.

2. Steeneken, P.A.M., Woortman, A.J.J., Gerritsen, A.H. and Poort, H., Powder Technol. 47(1986) 239.

3. Hollenbach, A.M., Peleg, M. and Rufner, R., J. Food Sci. 47(1982) 538.

4. Degussa Corp., Technical Bulletin Pigments, Pig. 31-5 -105-1084, 1984.

5. Visser, J., Powder Technol. 58(1989) 1.

6. Degussa Corp., Technical Bulletin Pigments, Pig. 56-2 -3-577K, 1977a.

7. Geldart, D., Powder Technol. 7(1973) 285.

8. Degussa Corp., Technical Bulletin Pigments, Pig. 62-2 -5-677K, 1977b.

9. Degussa Corp., Technical Bulletin Pigments, Pig. 11-7 -2-1084 DD e, 1982.

10. Geldart, D. and Wong, A.C.Y., Chem. Eng. Sci. 39(1984) 1481.

11. Wheeler, D.J., Understanding Industrial Experimentation, Statistical Process Controls Inc., Tennessee, 1988.

Table 1. Effect of AOC Addition on LPBD.

Run #	0.0% AOC	0.003125% AOC	0.00625% AOC
Run 1	789.60	802.00	842.60
Run 2	797.80	809.70	856.60
Run 3	795.50	811.10	861.40
Run 4	804.30	805.70	863.60
Run 5	795.10	800.70	847.60
Mean	796.46	805.84	854.36

Run #	0.0125% AOC	0.025% AOC	0.05% AOC
Run 1	885.70	954.0	996.10
Run 2	885.10	940.1	985.80
Run 3	882.30	934.5	982.30
Run 4	893.50	946.3	979.40
Run 5	884.50	939.1	976.50
Mean	886.22	942.8	984.02

Run #	0.1% AOC	0.2% AOC	0.35% AOC
Run 1	1041.20	1037.70	1069.10
Run 2	1027.10	1038.50	1054.40
Run 3	1034.30	1036.30	1061.40
Run 4	1026.20	1032.80	1048.00
Run 5	1017.30	1029.30	1069.90
Mean	1029.22	1034.92	1060.56

Run #	0.55% AOC	0.8% AOC	1.0% AOC
Run 1	1036.00	1050.70	999.8
Run 2	1035.70	1045.20	1012.1
Run 3	1030.30	1034.00	996.4
Run 4	1022.90	1038.00	993.5
Run 5	1028.70	1030.30	1001.7
Mean	1030.72	1039.64	1000.7

Run #	1.25% AOC
Run 1	1001.10
Run 2	990.70
Run 3	993.00
Run 4	1010.30
Run 5	1005.30
Mean	1000.08

Units of LPBD: kg/m3

Table 2. Effect of AOC Addition on TD

Run #	0.0% AOC	0.003125% AOC	0.00625% AOC
Run 1	1537.60	1526.40	1535.50
Run 2	1549.20	1541.50	1541.90
Run 3	1545.00	1526.30	1543.80
Run 4	1544.00	1538.80	1544.20
Run 5	1555.10	1541.40	1545.30
Mean	1546.18	1534.88	1542.14

Run #	0.0125% AOC	0.025% AOC	0.05% AOC
Run 1	1543.60	1593.00	1606.1
Run 2	1551.80	1593.50	1606.3
Run 3	1555.30	1593.40	1601.1
Run 4	1554.20	1590.40	1601.7
Run 5	1553.00	1596.60	1600.3
Mean	1551.58	1593.38	1603.1

Run #	0.1% AOC	0.2% AOC	0.35% AOC
Run 1	1618.00	1600.20	1610.50
Run 2	1615.40	1603.50	1603.40
Run 3	1611.90	1593.50	1600.00
Run 4	1607.20	1598.50	1585.70
Run 5	1606.20	1594.20	1597.70
Mean	1611.74	1597.98	1599.46

Run #	0.55% AOC	0.8% AOC	1.0% AOC
Run 1	1583.0	1574.20	1552.2
Run 2	1572.5	1567.60	1541.3
Run 3	1575.6	1567.10	1541.3
Run 4	1571.6	1567.20	1540.0
Run 5	1569.3	1559.30	1539.2
Mean	1574.4	1567.08	1542.8

Run #	1.25% AOC
Run 1	1532.30
Run 2	1535.50
Run 3	1531.50
Run 4	1528.60
Run 5	1524.70
Mean	1530.52

Units of TD: kg/m3

Table 3. Effect of AOC Addition of Hausner Ratio.

Run #	0.0% AOC	0.003125% AOC	0.00625% AOC
Run 1	1.947	1.903	1.822
Run 2	1.942	1.904	1.800
Run 3	1.942	1.882	1.792
Run 4	1.920	1.910	1.788
Run 5	1.956	1.925	1.823
Mean	1.941	1.905	1.805

Run #	0.0125% AOC	0.025% AOC	0.05% AOC
Run 1	1.743	1.670	1.612
Run 2	1.753	1.695	1.629
Run 3	1.763	1.705	1.630
Run 4	1.739	1.681	1.635
Run 5	1.756	1.700	1.639
Mean	1.751	1.690	1.629

Run #	0.1% AOC	0.2% AOC	0.35% AOC
Run 1	1.554	1.542	1.506
Run 2	1.573	1.544	1.521
Run 3	1.558	1.538	1.507
Run 4	1.566	1.548	1.513
Run 5	1.579	1.549	1.493
Mean	1.566	1.544	1.508

Run #	0.55% AOC	0.8% AOC	1.0% AOC
Run 1	1.528	1.498	1.552
Run 2	1.518	1.500	1.523
Run 3	1.529	1.515	1.547
Run 4	1.536	1.510	1.550
Run 5	1.525	1.513	1.537
Mean	1.527	1.507	1.542

Run #	1.25% AOC
Run 1	1.531
Run 2	1.550
Run 3	1.542
Run 4	1.513
Run 5	1.517
Mean	1.530

Table 4. Effect of A200 Addition on LPBD.

Run #	0.0% A200	0.003125% A200	0.00625% A200
Run 1	806.2	844.6	814.70
Run 2	795.5	834.5	819.70
Run 3	795.9	830.5	818.60
Run 4	796.3	841.3	808.80
Run 5	802.1	842.1	816.60
Mean	799.2	838.6	815.68

Run #	0.0125% A200	0.025% A200	0.05% A200
Run 1	895.50	937.20	991.90
Run 2	902.70	930.30	997.40
Run 3	889.40	935.10	1000.40
Run 4	894.40	935.00	1003.40
Run 5	893.10	944.60	996.20
Mean	895.02	936.44	997.87

Run #	0.1% A200	0.2% A200	0.35% A200
Run 1	1035.90	1054.9	1041.60
Run 2	1051.10	1051.0	1026.00
Run 3	1036.10	1051.2	1030.00
Run 4	1029.40	1056.3	1035.00
Run 5	1030.40	1056.1	1027.70
Mean	1036.58	1053.9	1032.06

Run #	0.55% A200	0.8% A200	1.0% A200
Run 1	993.30	974.1	930.4
Run 2	1000.80	976.6	921.5
Run 3	1005.80	979.1	919.3
Run 4	1006.30	980.2	930.6
Run 5	1003.90	980.5	933.2
Mean	1002.02	978.1	927.0

Run #	1.25% A200
Run 1	903.30
Run 2	905.40
Run 3	904.90
Run 4	900.90
Run 5	902.60
Mean	903.42

Units of LPBD: kg/m3

Table 5. Effect of A200 Addition on TD.

Run #	0.0% A200	0.003125% A200	0.00625% A200
Run 1	1541.20	1538.90	1528.50
Run 2	1555.80	1537.40	1541.60
Run 3	1558.70	1540.40	1530.10
Run 4	1557.30	1538.80	1531.70
Run 5	1559.30	1540.80	1537.50
Mean	1554.46	1539.26	1533.88

Run #	0.0125% A200	0.025% A200	0.05% A200
Run 1	1548.70	1595.40	1619.00
Run 2	1549.90	1589.30	1612.70
Run 3	1551.00	1591.90	1612.10
Run 4	1548.20	1586.50	1612.40
Run 5	1550.00	1590.20	1608.20
Mean	1549.56	1590.66	1612.88

Run #	0.1% A200	0.2% A200	0.035% A200
Run 1	1632.80	1628.40	1612.30
Run 2	1627.50	1622.60	1609.00
Run 3	1628.10	1623.00	1611.90
Run 4	1623.20	1622.20	1610.30
Run 5	1626.10	1619.70	1606.30
Mean	1627.54	1623.18	1609.96

Run #	0.55% A200	0.8% A200	1.0% A200
Run 1	1578.70	1544.40	1503.30
Run 2	1579.30	1551.00	1512.40
Run 3	1575.90	1547.50	1507.50
Run 4	1577.40	1548.00	1508.60
Run 5	1576.90	1546.20	1510.30
Mean	1577.64	1547.42	1508.42

Run #	1.25% A200
Run 1	1467.30
Run 2	1467.90
Run 3	1468.50
Run 4	1467.80
Run 5	1470.20
Mean	1468.34

Units of TD: kg/m3

Table 6. Effect of A200 Addition on Hausner Ratio.

Run #	0.0% A200	0.003125% A200	0.00625% A200
Run 1	1.912	1.822	1.876
Run 2	1.956	1.842	1.881
Run 3	1.958	1.855	1.869
Run 4	1.956	1.829	1.894
Run 5	1.944	1.830	1.883
Mean	1.945	1.836	1.881

Run #	0.0125% A200	0.025% A200	0.05% A200
Run 1	1.729	1.702	1.632
Run 2	1.717	1.708	1.617
Run 3	1.744	1.702	1.611
Run 4	1.731	1.697	1.607
Run 5	1.735	1.683	1.614
Mean	1.731	1.699	1.616

Run #	0.1% A200	0.2% A200	0.35% A200
Run 1	1.576	1.544	1.548
Run 2	1.548	1.544	1.568
Run 3	1.571	1.544	1.565
Run 4	1.577	1.536	1.556
Run 5	1.578	1.534	1.563
Mean	1.570	1.540	1.560

Run #	0.55% A200	0.8% A200	1.0% A200
Run 1	1.589	1.585	1.616
Run 2	1.578	1.588	1.641
Run 3	1.567	1.580	1.640
Run 4	1.567	1.579	1.621
Run 5	1.571	1.577	1.618
Mean	1.574	1.582	1.627

Run #	1.25% A200
Run 1	1.624
Run 2	1.621
Run 3	1.623
Run 4	1.629
Run 5	1.629
Mean	1.625

Table 7. Effect of Flow Conditioners on Fluid Bed Elutriation Losses.

0	1 No Additive	2 0.35% AOC	3 0.2% A200
1	20	17	10.9

Loss expressed as % of initial bed mass.

Table 8. Comparison of the Flow Conditioners.

	Cohesive Powder	Powder + AOC	Powder + A200
Optimum loading (%)	--	0.35	0.2
Max bed expansion (%)	27.0	90.0	91.0
Max bed pressure drop, % of ideal value	87.8	90.3	97.5
Bed pressure drop at max flow, % of ideal	85.0	83.1	92.8
Elutriation loss, % of initial bed mass	20.0	17.0	10.9

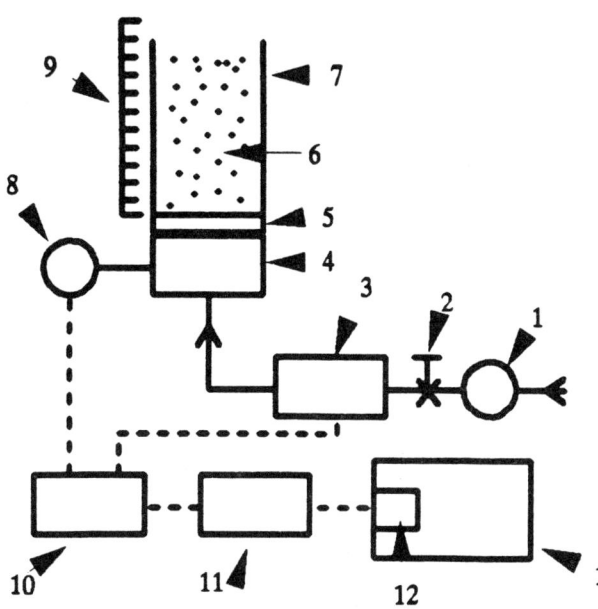

Figure 1. Fluid Bed Experimental Setup.

1: pressure regulator
2: plug valve
3: 50 SLM(0.833E-3 m³/s) mass flow controller
4: plenum
5: distributor plate
6: powder
7: fluid bed glass wall
8: pressure transducer, 50" H$_2$O g (12,291.2 Pa g)
9: bed height scale
10: screw terminal board, STA 08
11: expansion multiplexer, EXP16
12: analog to digital board, DAS8
13: IBM 7531 industrial computer
--- electrical line

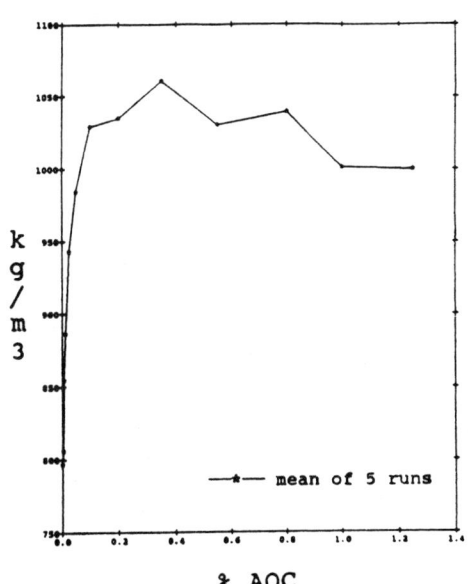

Figure 2. Effect of AOC Addition on LPBD.

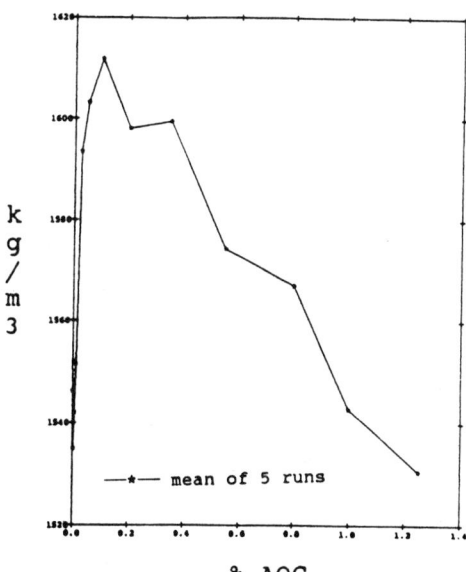

Figure 3. Effect of AOC Addition on TD.

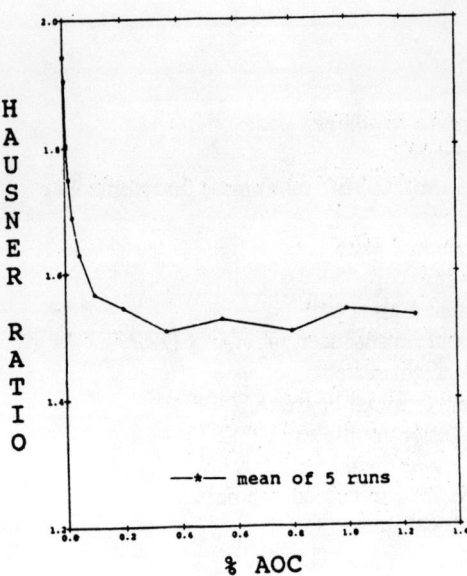

Figure 4. Effect of AOC Addition on Hausner Ratio.

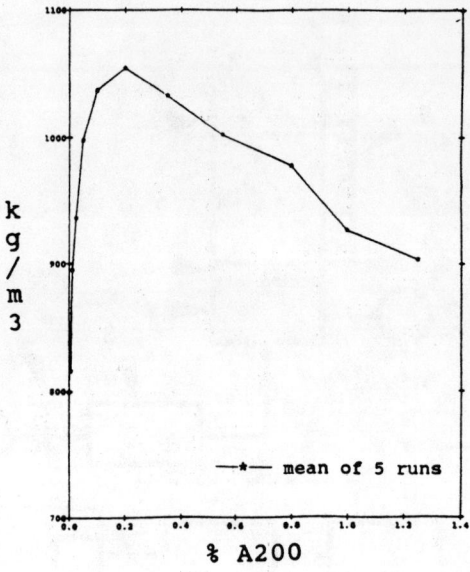

Figure 5. Effect of A200 Addition on LPBD.

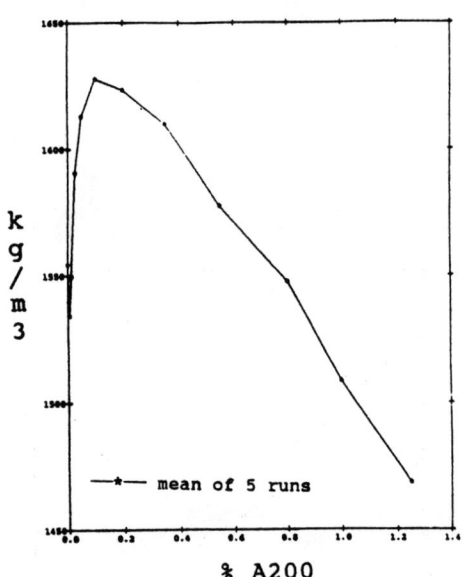

Figure 6. Effect of A200 Addition on TD.

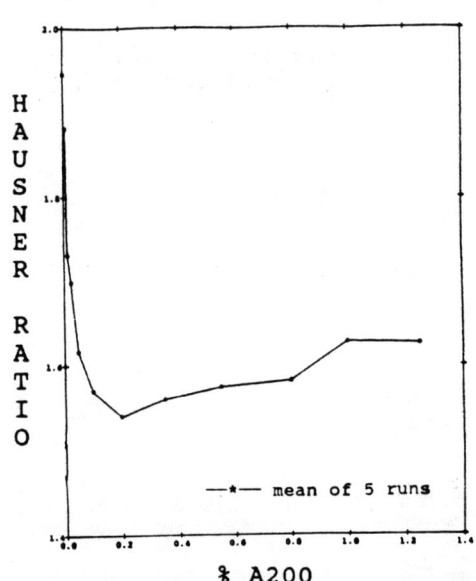

Figure 7. Effect of A200 Addition on Hausner Ratio.

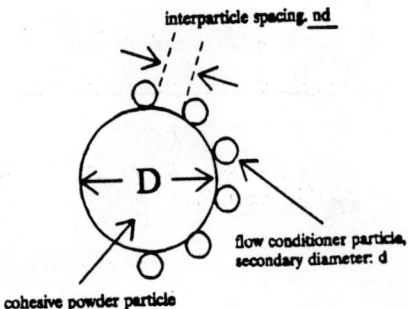

Figure 8. Flow Conditioner Loading.

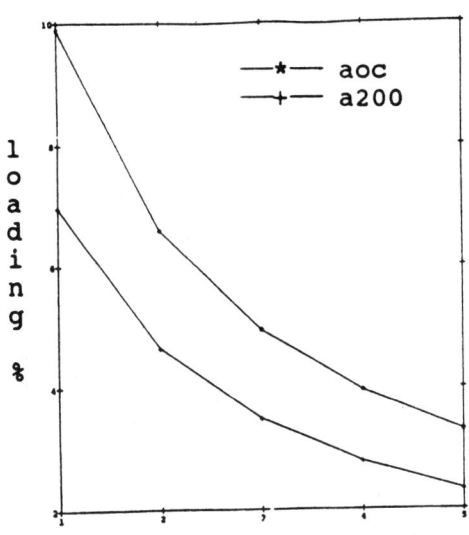

Figure 9. Effect of Flow Conditioner Interparticle Spacing on Loading of Same.

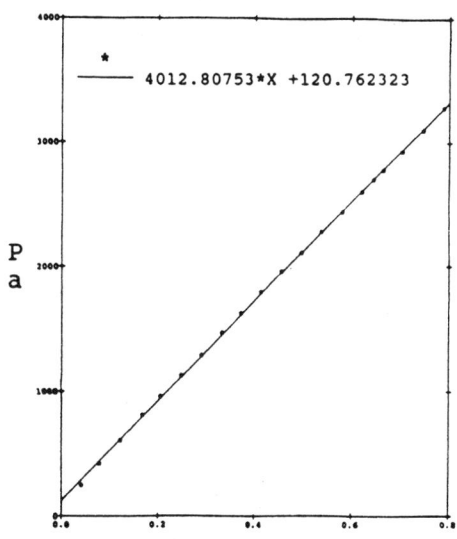

Figure 10. Distributor Pressure Drop vs Air Flow Rate.

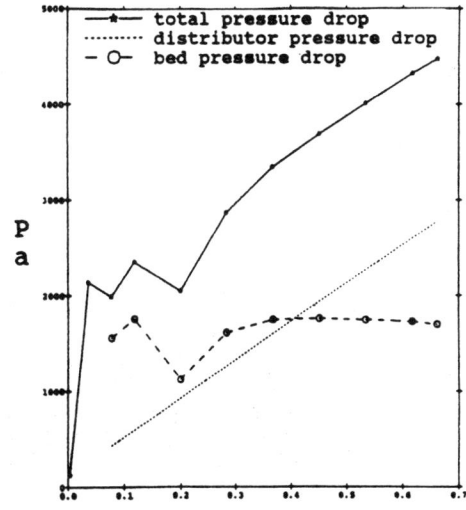

Figure 11. Mean Pressure Drop vs Mean Air Flow Rate Fluidization with No Fluid Aid.

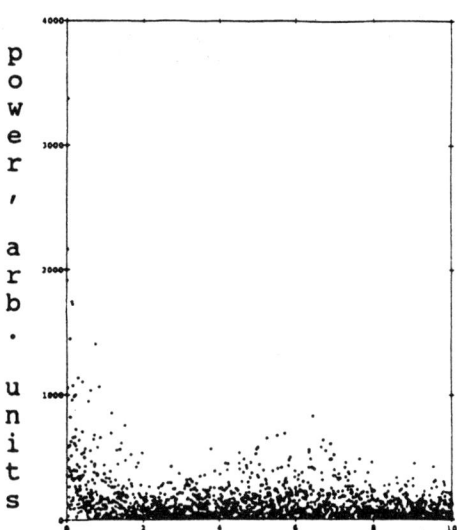

Figure 12. Power Spectrum of FFT of Plenum Pressure Fluidization with No Flow Conditioner.

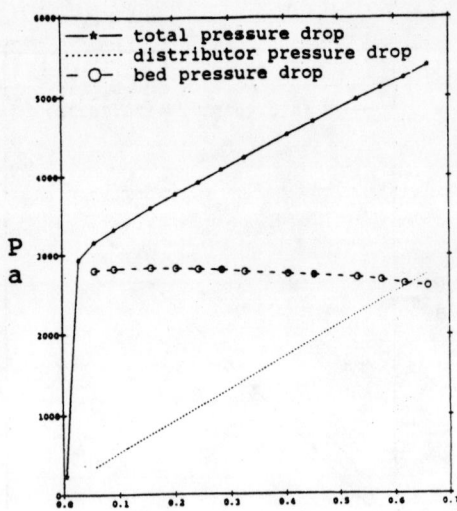

Figure 13. Mean Pressure Drop vs Mean Air Flow Rate Fluidization with the Addition of 0.35% AOC.

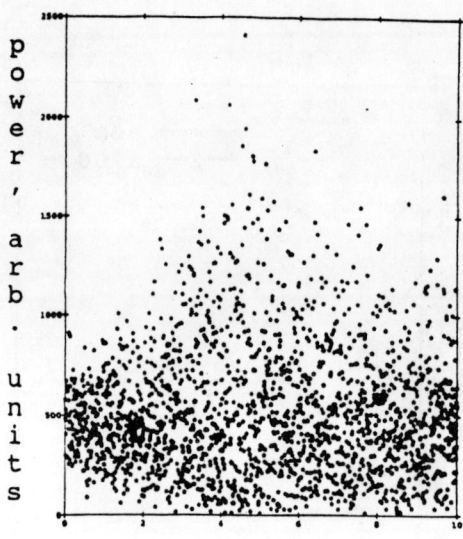

Figure 14. Power Spectrum of FFT of Plenum Pressure Fluidization with the Addition of 0.35% AOC.

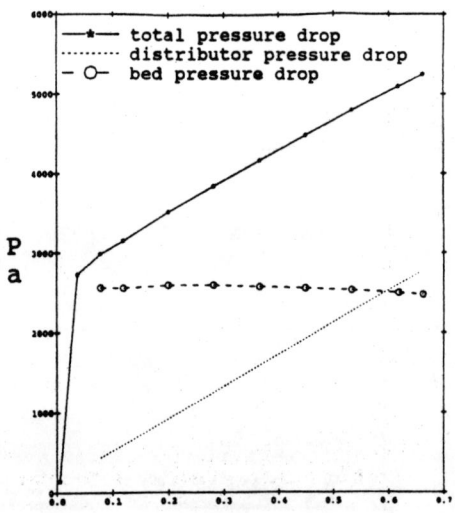

Figure 15. Mean Pressure Drop vs Mean Air Flow Rate Fluidization with the Addition of 0.2% A200.

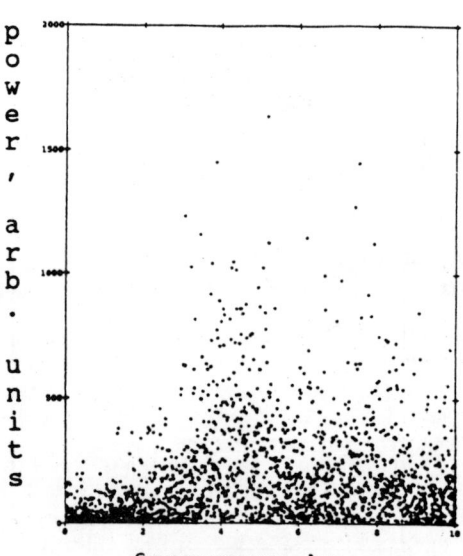

Figure 16. Power Spectrum pf FFT of Plenum Pressure Fluidization with the Addition of 0.2% A200.

PRELIMINARY CAPACITANCE IMAGING EXPERIMENTS OF A FLUIDIZED BED

J.S. Halow and G.E. Fasching ■ U.S. Department of Energy, Morgantown, WV
P. Nicoletti ■ EG&G Washington Analytical Services Center Inc., Morgantown, WV

A unique rapid imaging system based on the measurement of capacitance is being developed at the Morgantown Energy Technology Center (METC). This paper presents the results of preliminary experiments in which a fluidized bed was imaged with the system. Capacitance imaging permitted high speed non-intrusive maps of the voidage distribution within the fluidized bed to be obtained. These maps show the rise of slugs and bubbles in the bed in slow motion. Quantitative data derived from the mapping includes bubble and slug sizes, rise velocities, spacing and frequency. The data also illustrate effects such as bubble coalescence and other phenomena. Three dimensional contour images of slugs have also been constructed from the imaging of a coarse particle bed. A system with increased radial resolution and an increased number of axial imaging planes is currently being developed.

METC has an active development program in a number of fossil energy technologies including fluidized bed combustion, coal gasification, hot gas cleanup and oil shale retorting. Many of the systems conceived for these technologies include fluidized beds as processing reactors or contactors. Fossil energy fluidized beds are typically coarse particle systems. The fluidization behavior of these coarse particle systems have not been extensively studied and fluidization data and models to describe them are limited. Capacitance imaging was chosen as a method to develop the detailed information needed to define correlations and models to predict performance of these beds.

Capacitance measurements to determine point values of voidage in fluidized beds have been used since the 1960s in a variety of studies reported in the literature. Various techniques have evolved which use one or more sets of capacitance sensors within the bed or mounted external to the bed to determine point or average voidages. By measuring time shifts in capacitance traces separated vertically, bubble velocity can be determined. In the 1970s, work on capacitance systems at METC was initiated to detect bed levels, measure solids flow rates and other quantities of interest in fossil energy processes. This work developed to non-intrusive wall-mounted electrodes measuring average voidages through an axial segment of a bed. The current system simultaneously measures voidages in separate volumes within an axial segment. The system used in this study measured capacitance in a cross-section divided into 49 volumes or "pixels" at four adjacent vertical levels. This paper reports the results of a preliminary study with this system.

DESCRIPTION OF CAPACITANCE IMAGING AND FLUIDIZATION SYSTEM

The imaging system is incorporated into a 15.24 cm diameter fluidized bed operated at ambient temperature and pressure. The bed contains four rings of sensing electrodes mounted flush with the walls of the fluidized bed. Each ring of electrodes consists of 16 individual sensing electrodes spaced symmetrically on the interior circumference of the bed. These sensing electrodes are energized in pairs and the current along the flux tube between them is measured and stored. Rapid switching of the energized pairs generates a set of data related to the dielectric constant of the material along the flux tubes. Other pairs of electrodes are then energized which provided flux tubes that intersect the first set of flux tubes. The dielectric constant within individual volumes can be calculated from the combined data sets. This procedure is repeated rapidly at each of the four levels to provide the measurements with time. Above and below the sensing electrodes are guard electrodes which are energized in such a way as to confine and shape the flux tubes between energized sensing electrodes. Figure 1 illustrates the bed and the principal parts of the imaging system.

The 49 pixels created by the intersection of the flux tubes at each sensing level are illustrated in Figure 2. Measurements are made by applying a positive voltage sequentially to electrodes 2 through 8 in Figure 2 and a negative voltage to electrodes 10 through 16. Figure 3 illustrates the flux tubes generated by the sequential application of the voltages. Additional measurements are obtained by shifting circumferentially the electrodes energized. The sensing procedure results in 104 measurements for each time interval at each of the four levels in the bed. Since the measured currents result from the dielectric constant of the material within each of the pixels that the flux tube crosses, a set of 104 simultaneous equations can be written and solved to

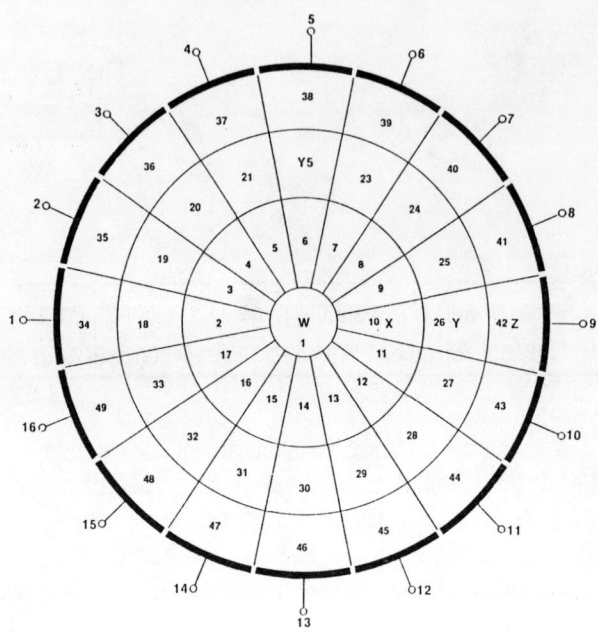

Figure 2. Pixel definition.

determine the capacitance within each pixel. These measurements of capacitance are assumed to be proportional to the volume fraction of solids in the pixel. Voidage is the volume fraction of air or 1 minus the volume fraction of solids. In the current experiments these measurements were repeated 100 times per second to provide a

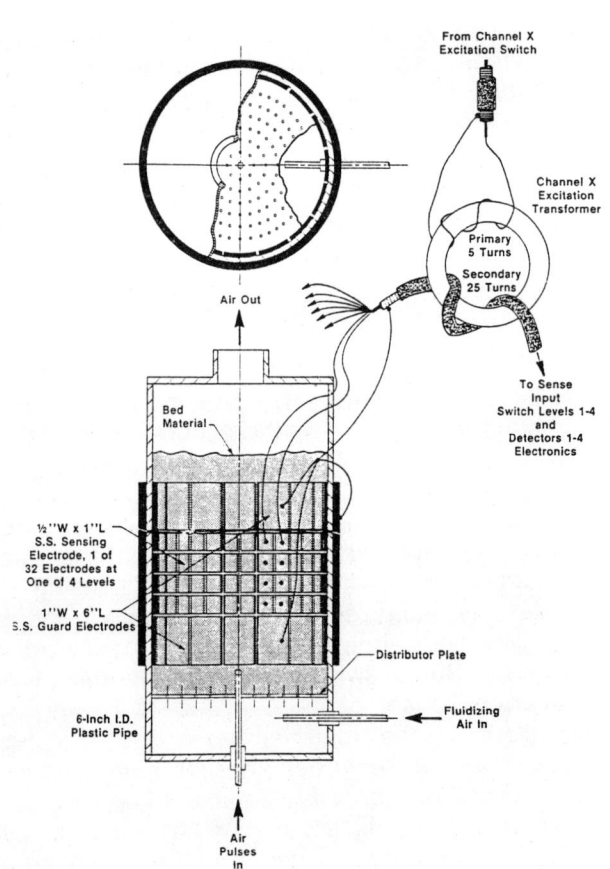

Figure 1. Capacitance imaging system.

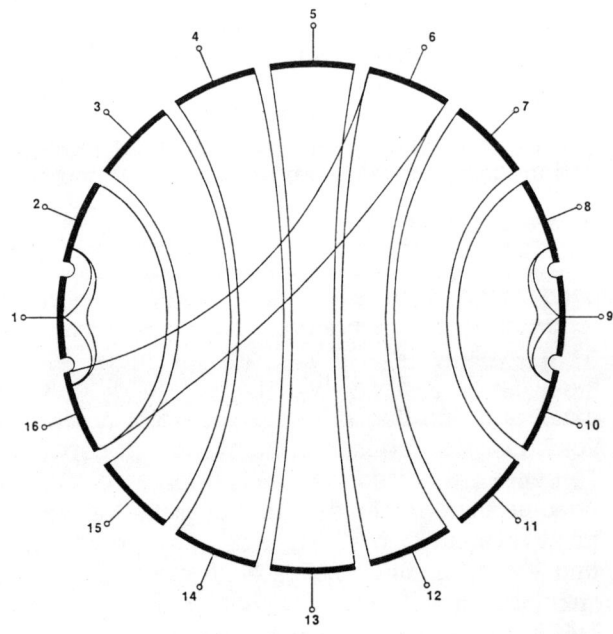

Figure 3. Flux tubes in imaging system.

temporal resolution of 0.01 seconds. A more complete description of the imaging system and the techniques used is provided in Fasching and Smith (1).

Air was provided to the bed through a grid containing 169 holes (0.2 cm diameter) spaced on an equal area basis. The total open area of the grid was 2.9% and the grid pressure drop varied from 0.3 to 12 times the bed pressure drop. With the fine catalyst bed material, a fine screen was attached below the grid to reduce weepage and increase grid pressure drop. The grid was located 20.5 cm below the first level of sensing electrodes. Each sensing electrode was 2.33 cm high and were spaced 2.54 cm vertically apart. All of the holes except the center hole were supplied air from a plenum. The center hole was equipped with a separate air supply but was not used in this study. A high speed data acquisition system was used to measure the air flow rate to the plenum, the pressure in the plenum and the pressure 2.54 cm above the grid.

DESCRIPTION OF EXPERIMENTS

The experiments reported in this paper were preformed to test the imaging system and to provide measurements to develop data analysis techniques. Particle systems and fluidization conditions were chosen for which existing information and correlations might be expected to apply. Three types of particles were used in the experiments including 1/8 inch nylon spheres, 700 micron irregularly shaped plastic particles, and a coarse size cut of a commercial fluid cat cracking catalyst. Table 1 provides the properties of these particles and the fluidizing conditions. Fluidization velocities were chosen starting at or slightly above minimum fluidization and progressing up to several times minimum fluidization.

At the beginning of each experiment the bed was vibrated for ten seconds to bring it to a known packed state at which a baseline set of measurements were made to calibrate the imaging system. Then fluidizing air was introduced and, after several seconds of fluidization to "unpack" the bed, the imaging was started. Measurements continued for 10 seconds with data acquired at each level at 100 times per second.

A real time analysis of the data is performed by the instrument using an approximate method for simultaneous solution of the 104 equations and the results displayed on a monitor. This aids in the experiments and is also videotaped for later review. More accurate conversion of the current measurements along the flux tubes to voidages is performed later from the stored data. A total of 196,000 values of voidage are determined during each 10 second experiment. Ultimately the data is displayed on a micro computer as four ellipses stacked one above another representing the four sensing levels in the bed. Each ellipse is divided into 49 pixels representing the measured volumes and the voidage in each volume illustrated by filling each pixel with either a color or gray scale tone from a palette representing voidages ranging from the compact state to an all-air condition. The computer successively colors the pixels with data from each time measurement rapidly enough to provide a slow motion version of the voidage variations with time.

Figure 4 illustrates the voidage distributions at the four sensing levels at 0.65, 0.85, and 1.05

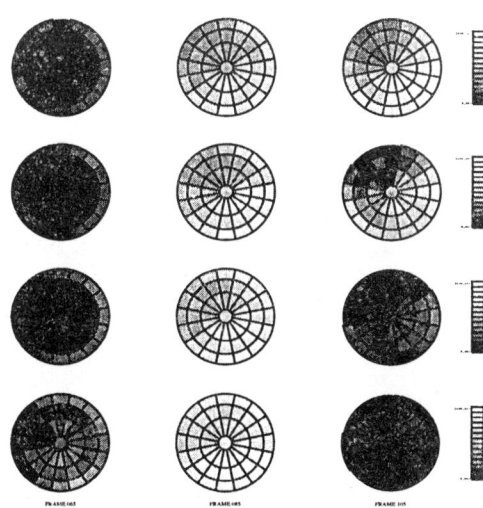

Figure 4. Computer imaging output.

Table 1.
Particle Properties

Material	Weight Average Diameter Microns	Particle Density gm/cc	Bulk Density gm/cc	Packed Voidage	Minimum Fluidization Velocity cm/sec
1/8-Inch Nylon Spheres	3175	1.12	0.64	0.40	84.1
700-Micron Plastic	704	1.55	0.73	0.53	19.0
Cracking Catalyst	70	1.44	0.63	0.38	0.70

seconds into an experiment. Lighter shades indicate higher voidages while darker shades lower values of bed voidage. The four imaging levels are represented by the four vertically stacked ellipses, with the bottom ellipse representing the lowest imaging level. In this sequence a large slug is passing through the imaging fields.

Further processing of the data was performed to determine bubble and slug properties. To do this some bubble shape must be assumed. Real bubbles have shapes resembling geometric shapes but are not precisely described by these assumed shapes. As a first choice to perform this analysis, average voidage versus time data was generated by weight averaging the individual pixel data based on pixel area. Determination of bubble and slug sizes and velocities from these averages will be described below. Reviewing the movie-like computer images greatly aided in the interpretation of the average voidage plots. Other techniques for determining bubble properties from the complete data sets are being examined.

In the following analysis, experimentally determined bubble sizes and velocities are correlated in various ways assuming that bubble theory relationships, such as in Davidson and Harrison (2), between size and velocity would be applicable. These relationships are summarized below:

for free bubbles

$$U_a = U_o - U_{mf} + 0.71 \sqrt{gD_s} \qquad (1)$$

for slugs

$$U_a = U_o - U_{mf} + 0.35 \sqrt{gD_t} \qquad (2)$$

for wall slugs

$$U_a = U_o - U_{mf} + 0.49 \sqrt{gD_t} \qquad (3)$$

In these relationships U_a is the rise velocity of the bubble or slug, U_o is the superficial gas velocity through the bed, U_{mf} is the minimum fluidization velocity, D_s is the diameter of a sphere with a volume equal to the bubble, D_t is the tube diameter and g is the acceleration of gravity.

RESULTS FOR 1/8 INCH-NYLON SPHERES

A total of seven imaging experiments were conducted with the coarse particle 1/8-inch nylon spheres. Superficial bed velocities ranged from 87.2 cm/sec (slightly above the minimum fluidization velocity of 84.1 cm/sec) to 126 cm/sec or 1.50 times minimum fluidization velocity. Initially spheres were placed in the bed to provide a slumped bed high of 65 cm. At higher velocities some of the spheres were removed to prevent overflowing the vessel due to the expansion of the bed. Visual observation of the bed surface indicated that over most of the velocity range, the bed exhibited a slugging behavior. At the lower velocities the slugs appeared to be attached to the wall while at higher velocities they filled most of the bed cross-section.

Figure 5 illustrates the average cross-section voidages from the experiment at the lowest superficial velocity. The four curves on the figure are for the four sensing levels. The figure shows the passage of a slug through the imaging system by the successive rise and fall in average voidage at each level. The peaks do not always rise to the same maximum indicating different maximum cross-sections. Although the peaks are periodic they do vary from a precise frequency. An interesting feature of these data is the slight compression of the bed immediately preceding several of the bubbles. This feature probably indicates a defluidized condition in which spheres in contact with each other are being pushed upward by the approaching bubble. This compression was not evident in the higher velocity experiments.

Figure 6 is an expanded scale plot of the data for the bubble which passed between 2.5 and 3.0 seconds into the experiment. The shift in the voidage pulse at the four levels is clearly

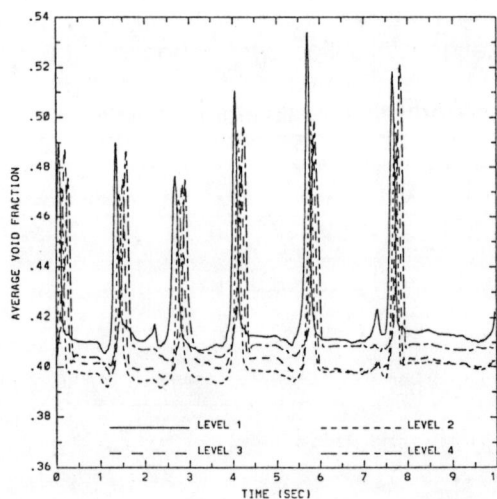

Figure 5. Nylon spheres—average voidage at low superficial velocity.

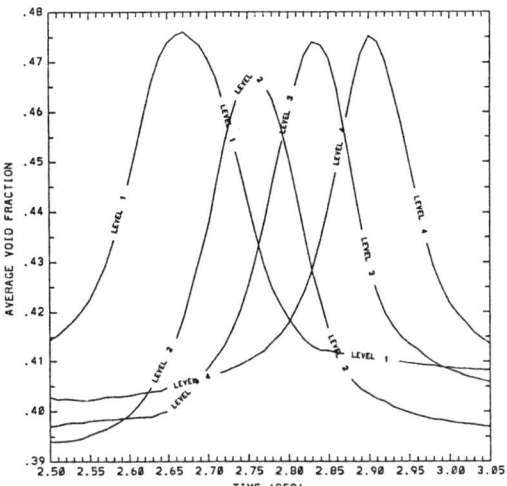

Figure 6. Nylon spheres—average voidage at low superficial velocity—expanded scale.

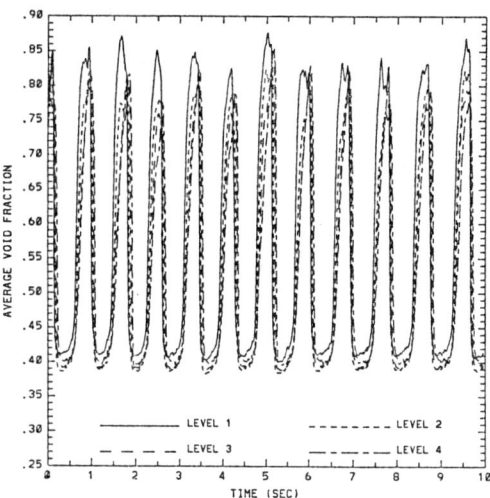

Figure 7. Inch nylon spheres—average voidage at high superficial velocity.

illustrated. The velocity of the slug was determined from the time shift of the down side of the peak. This represents the passage of the end of the bubble through each imaging level. The volume of the bubble at each level was determined by integrating the data to obtain the area under each voidage peak. A peak voidage was also determined for each bubble. A maximum diameter assuming a circular cross-section was calculated from these peaks while a length was calculated from this dimension and the volume assuming the bubble had a hemispherical cap attached to a cylindrical body. Images of the voidages at low superficial velocities indicated that they were round-nosed wall slugs. Data at higher velocity showed that these voidages were round-nosed slugs with diameters nearly as large as the tube and with lengths exceeding the tube diameter.

Figure 7 is the average voidage plot for the experiment at the highest superficial velocity. Peaks for the velocity are much higher and broader. Figure 8 shows the expanded scale voidage plot for the slug between 0 and 0.6 seconds. This slug had an equivalent maximum diameter of 12.8 cm and a length of 28.2 cm. It was travelling at a velocity of 91 cm/sec. Figure 4 shows three of the voidage distributions for this slug. At 0.65 seconds, the slug is entering the lowest imaging level. Note that an increase in voidage along the wall has preceded the slug and is shown at levels 2-4 at 0.65 seconds. This wall voidage increase occurred frequently with slugs in this set of experiments. At 0.85 seconds, the slug occupies all four voidage levels. At 1.05 seconds, it shows partially in levels 3 and 4.

A three-dimensional representation of this slug is shown in Figure 9. This figure was constructed by drawing contour lines from the voidage data at level 4 at successive times and plotting these vertically with the spacing determined by the slug velocity. The contours were drawn at a voidage of 0.66 chosen (somewhat arbitrarily) to mark the "boundary" of the slug. A depression near the tail of the slug is evident.

Figure 10 is a plot of the bubble/slug velocity versus the bubble/slug volume for the seven experiments. There is a wide variation in both the velocity and size of gas voids at each superficial velocity. In these experiments, the imaging

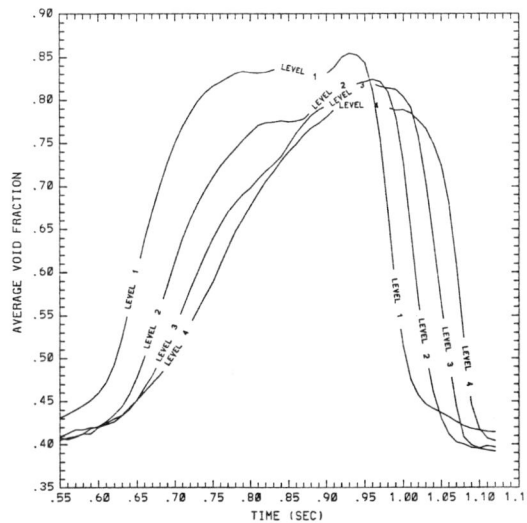

Figure 8. Nylon spheres—average voidage at high superficial velocity—expanded scale.

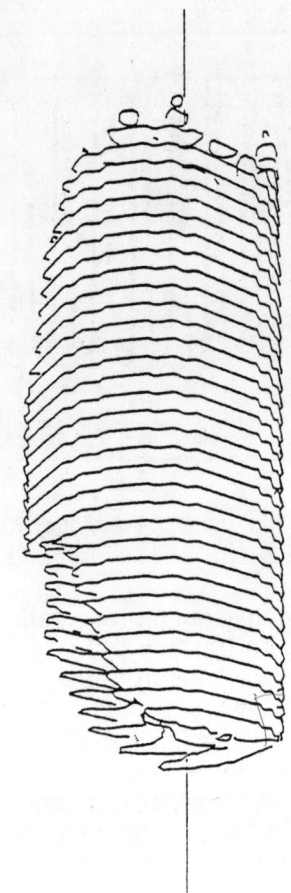

Figure 9. Nylon spheres—image of slug.

levels were from 1 to 1.7 diameters above the grid. Slugs generally would not be expected at this low height to diameter ratio.

Slug growth however appear to be rapid in this coarse material. Slugs also appear to still be growing when they passed the imaging levels. Despite the variability in the slug velocities in each experiment, there is a consistent relationship between slug volume and rise velocity.

Averages of the bubble properties in all of the experiments for the various superficial velocities is presented in Table 2. In this table D_c is the diameter based on the peak cross-sectional voidage, L_s is the length of a hemispherically capped cylindrical slug calculated from D_c and the measured slug volume, V is the measured volume, and S_p is the distance between the base of a bubble or slug and the base of the preceding bubble assuming the preceding bubble continued to travel upward at the measured rise velocity. The determination of the properties for the plastic and catalyst is explained below.

The average of the single slug rise velocities at each superficial velocity generally satisfies equation (2) for slug velocities greater than 100 cm/sec. Although the average of the slug velocities in an experiment gives the expected behavior, a similar plot using the individual slug velocities gives a plot with the expected trend but with a high degree of scatter and does not provide a good method of correlating this data. This is a result of the variability of slug velocity at a given superficial velocity. A much better correlation is achieved by plotting the actual slug rise velocity versus a measure of the slug size. Figure 11 shows a plot of actual rise velocity versus slug

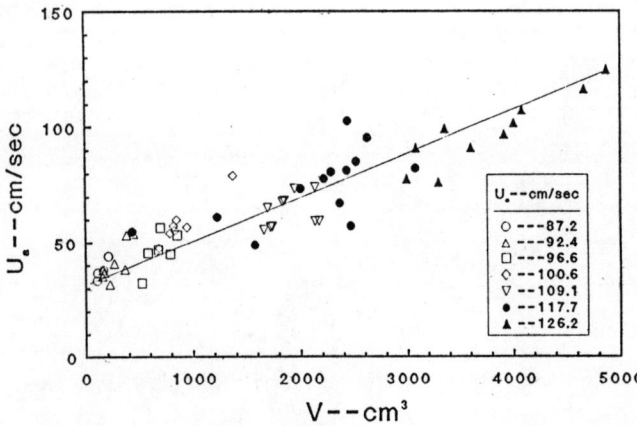

Figure 10. Nylon spheres—velocity versus slug volume.

Table 2. Average Bubble Properties

Material	Ave U_o cm/sec	Ave U_a cm/sec	Ave D_c cm	Ave D_s cm	Ave L_s cm	Ave V cm³	Ave S_p cm	Frequency sec-1
1/8-Inch Nylon Spheres	87.2	39.0	5.79	6.63	6.83	167.3	--	0.67
	92.4	41.1	7.26	7.87	7.42	268	--	0.84
	96.6	46.6	10.52	10.92	9.70	688	--	0.70
	100.6	58.5	10.29	11.91	12.47	900	--	0.80
	109.1	63.7	12.32	15.32	17.96	1889	--	1.00
	117.6	74.4	11.99	15.70	20.3	2129	--	1.34
	126.2	100.6	12.60	19.46	33.4	4642	--	1.18
700-Micron Plastic	19.1	35.3	--	2.67	--	--	11.9	2.50
	20.2	42.8	--	3.09	--	--	10.4	3.88
	21.7	47.7	--	3.44	--	--	11.6	4.00
	23.3	46.5	--	4.27	--	--	12.2	3.60
	25.9	53.4	--	5.29	--	--	14.8	3.60
	28.5	64.5	--	5.65	--	--	14.5	4.00
	31.6	77.4	--	6.21	--	--	16.9	4.10
Cracking Catalyst	0.70	30.4	--	2.30	--	--	8.6	3.60
	0.93	43.7	--	3.66	--	--	6.4	6.70
	1.24	55.3	--	5.00	--	--	7.2	7.72
	1.55	59.8	--	5.83	--	--	6.6	8.80

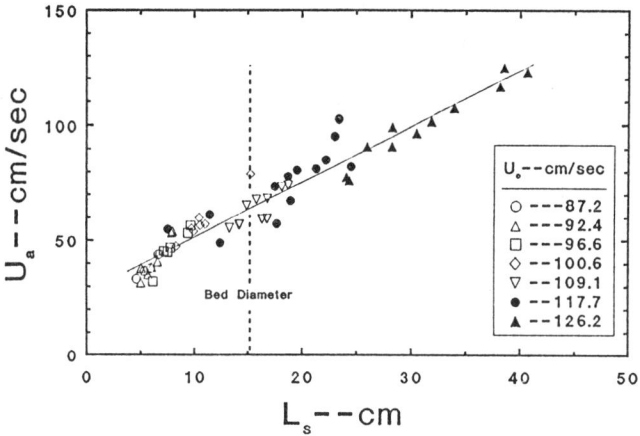

Figure 11. Nylon spheres—slug velocity versus slug length.

length. For wall slugs in the lower velocity experiments with diameters somewhat smaller than the tube size, the slug length was taken equal to the slug diameter. The break in the curve corresponds to where the slug has grown to approximately 80% of the tube diameter. Beyond this additional gas goes into increasing the slug length without an increase in diameter. Figure 12 shows a plot of the equivalent cylindrical diameter based on the peak voidage versus the slug length calculated from the volume and cylindrical diameter. The figure illustrated that the length and diameter are approximately equal up to about 80% of the tube diameter. Beyond that slugs grow mainly by increasing their length without much increase in diameter.

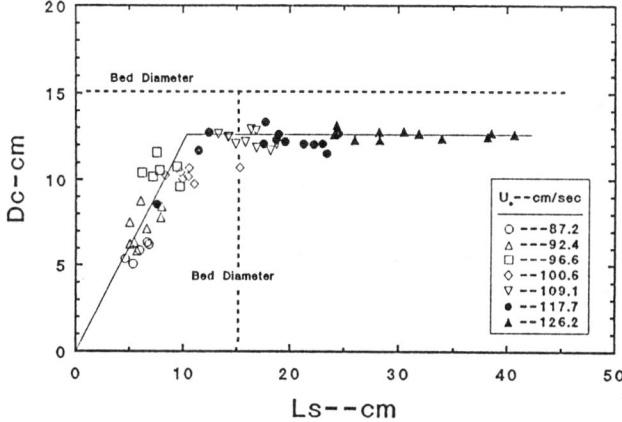

Figure 12. Nylon spheres—slug diameter versus slug length.

RESULTS FOR 700 MICRON PLASTIC

A total of seven imaging experiments were conducted with the 700-micron plastic particles. Superficial velocities ranged from 19.1 to 35.7 cm/sec. From the imaging data, average cross-sectional voidage versus time plots were determined. Unlike the data from the 1/8 inch spheres these plots did not show regular patterns of bubbles. The voidage plots showed the expected time shift with level in the bed but bubbles were variable in size and in their spacing. Frequently voidage peaks would overlap indicating the presence of more than one bubble within the 2.54 cm thickness of a sensing level. Because of this overlap it was not feasible to calculate bubble volumes by determining the area under the voidage rises.

Peak voidages were, however, determined and from these the diameter of the bubble could be calculated. When peaks overlap however, an error is introduced because of the inclusion of voidage rise due to the overlapping peak in the peak voidage determination. From the peak voidage, a bubble volume can be calculated if a bubble shape is assumed. In analyzing the 700 micron plastic and the cracking catalyst average voidage plots, a spherical bubble was assumed. As a bubble of this shape rises through the 2.54 cm thick imaging cross-section, the average voidage will rise reaching a maximum when the base of the bubble is at the mid-line of the cross-section. If the bubble diameter is less than 2.54 cm, the peak voidage is related to the volume of the spherical bubble. If the diameter is larger than 2.54 cm, the peak voidage volume is related to the volume of the segment of the bubble contained in the imaging segment. In both cases, the bubble diameter can be calculated from the measured peak voidage. From the voidage plots at successive levels, time shifts in the peaks were determined and from these the bubble rise velocity calculated. Overall from the experiments, the bubbles ranged from relatively small bubbles in the low velocity experiments to wall slugs and full slugs in the high velocity experiments.

The experiment at 19.1 cm/sec was at a condition that appeared to be minimum fluidization. The average cross-sectional voidage versus time plot is shown in Figure 13. Initially through the first three seconds of the experiment there were no discernable bubbles; there were, however, some minor voidage fluctuations. Between four and five seconds into the experiment, bubbling

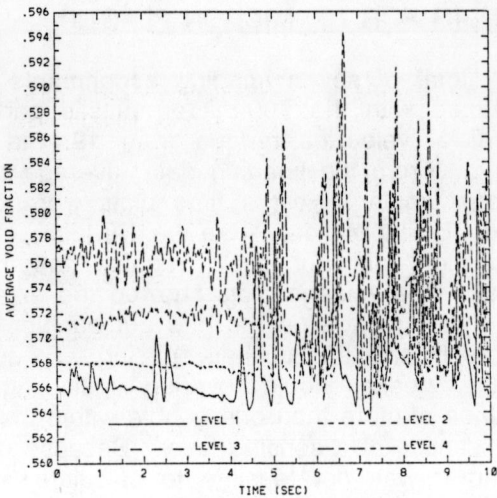

Figure 13. 700 micron plastic—start of fluidization.

started. Bubble diameters calculated from the plots varied between 2.3 to 3.1 cm while the velocities varied from 19 to 48 cm/sec. There were no discernible changes in the bed voidage at the onset of bubbling other than the start of bubbling itself.

Actual rise velocity versus bubble diameter for all seven of the imaging experiments is shown in Figure 14. A general trend is shown in the plot. Bubble velocity initially increases with bubble diameter and then levels off. This indicates a change in flow regime from free bubbles to slugs. Notably there are a substantial number of bubbles with velocities much higher than this general trend. From the voidage plots, the spacing between adjacent peaks can be determined as a time difference and from this a vertical spacing can be calculated assuming the higher bubble continues to rise at the same velocity it had when it passed the imaging levels. It was found from this information that most of the abnormally high velocity bubbles in Figure 14 were bubbles closely spaced behind the next bubble above them. They appear to be bubbles rising in the wakes of the upper bubbles. Note that with the small bubbles in this experiment, it is possible that closely vertically spaced bubbles may not necessarily be in line, i.e. they could be more or less side by side, in which case the trailing bubble should not show an abnormally high rise velocity. Many bubbles exhibit this kind of behavior.

Figure 15 shows the effect of bubbles rising in leading bubble wakes. This figure plots $(U_a - U_o + U_{mf})$ divided by $\sqrt{gD_t}$ against spacing divided by the diameter of the leading bubble. The velocity ratio should equal approximately 0.35 for slugs and 0.49 for wall slugs. The velocity ratio is plotted against the bubble spacing (i.e., the distance between the trailing edge of the bubble and the bubble nearest it higher in the bed divided by the diameter of the higher bubble. The plot shows that most points fall in a region expected for full slugs and wall slugs but most of the high velocity bubbles are between one to two leading-bubble diameters behind the lead bubble. If a bubble wake is approximately equal to the bubble diameter, the trailing bubble would enter the leading bubble's wake at about 1.5 (if the two bubbles were of equal size). At low superficial velocities some of the points fall below the others points indicating that the velocity scaling to a slugging bed is not correct. A scaling to the quantity $\sqrt{gD_s}$ would be more appropriate.

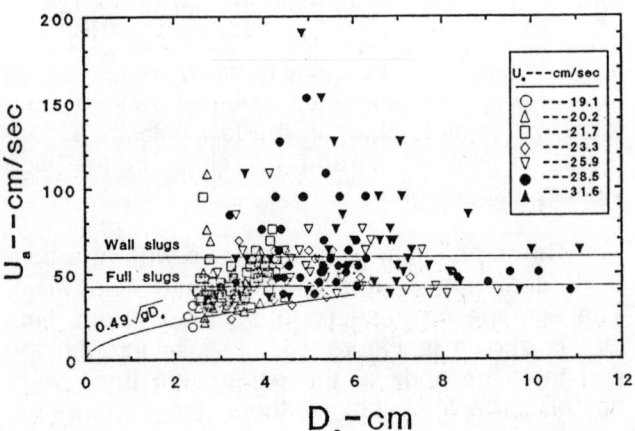

Figure 14. 700 micron plastic—individual bubble velocities.

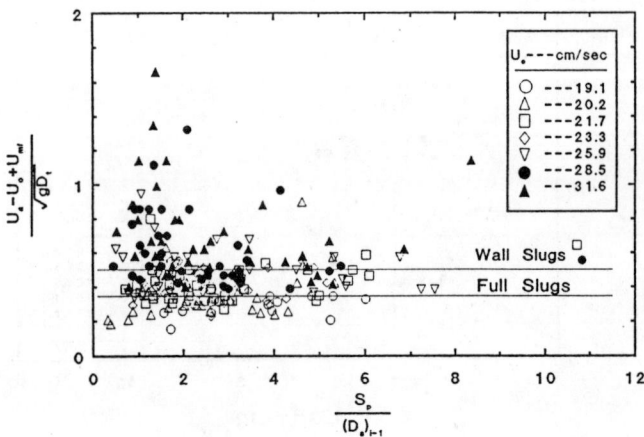

Figure 15. 700 micron plastic—bubble velocity dependence on bubble spacing.

RESULTS FOR CRACKING CATALYST

A total of eight experiments were conducted with the fine 70 micron cracking catalyst. Superficial velocities covered a range of from 0.7 cm/sec to 3.1 cm/sec. This material has a minimum fluidization velocity of approximately 0.7 cm/sec. The imaging data indicated a large number of small bubbles generated with this material. However, because these bubbles appeared to be of sizes approaching the size of the imaging pixels, no indications of shape were revealed. Average cross-sectional voidage plots versus time were also constructed for these experiments. These plots were similar to those obtained with the 700 micron plastic material showing peaks shifted in time with levels. There were, however, far more bubbles present and substantially more overlapping. Because of this it was only possible to determine bubble properties for the four lowest superficial velocity experiments. Equivalent hemispherical bubble diameters, actual rise velocities and spacing to the nearest higher neighbor were determined as was done with the 700-micron plastic experiments.

Figure 16 presents a plot of actual rise velocity versus diameter for the four experiments. The velocity shows an increase with increasing diameter over the range of 2 to 5 cm in diameter. Like the plastic data, there are a number of bubbles which show higher than expected velocities. Figure 17 shows a plot of $(U_a - U_o + U_{mf})$

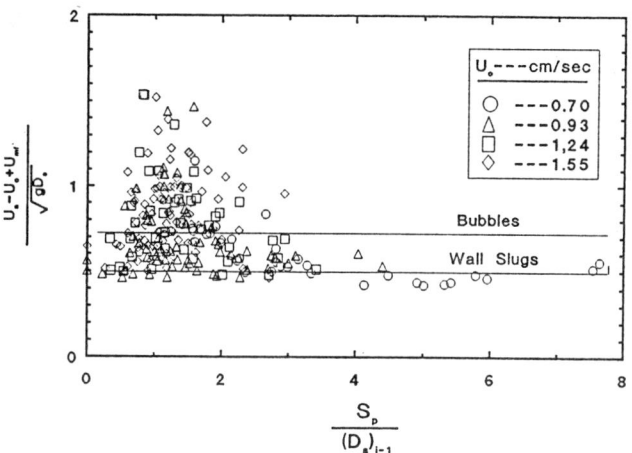

Figure 17. Cracking catalyst—bubble velocity dependence on bubble spacing.

divided by $\sqrt{gD_s}$ against spacing divided by lead bubble diameter. This scaling would be appropriate for a freely bubbling system. A band of points appears between 0.5 and 0.7 expected for wall slags and bubbles. As with the plastic data, most of the high velocity points are contained within a bubble spacing from about 0.5 to 2.5 bubble diameters from the leading bubble.

With the cracking catalyst, some expansion of the emulsion phase was observed from the voidage measurements in pixels without bubbles present. This expansion increased rapidly with increasing superficial velocity from 0.45 at minimum fluidization to 0.517 at two times minimum fluidization velocity after which it declined slightly. A more modest degree of expansion was observed with the 700 micron plastic. With the nylon spheres, no expansion of the emulsion phase was observed.

CONCLUSIONS

The preliminary experiments with a unique capacitance imaging system have given results generally consistent with expected rise velocity and size relationships. The generation of an average cross-sectional voidage versus time data from the total imaging data was successful for quantifying bubble properties it cases where bubble density is not high. Imaging of two smaller particle systems indicated that bubble rise velocities follow generally expected average trends but with a portion of the bubbles travelling at higher velocities for some bubbles where spacing between bubbles is less

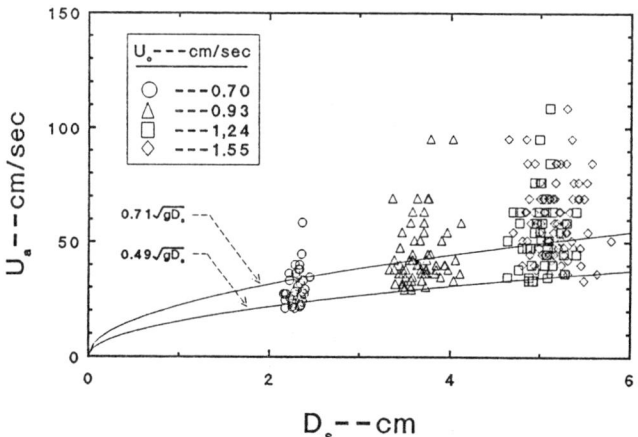

Figure 16. Cracking catalyst—individual bubble velocities.

than about 2 to 2.5 bubble diameters. From all of the data, it is clear that under a given fluidization condition, a distribution of bubble or slug sizes existed in these experiments. The average of the rise velocities slugs observed with a coarse material agrees well with the calculated velocity based on the tube diameter but individual slugs vary in size and do not follow this model. Slug rise velocity does correlate well with the slug length.

Currently the imaging system is being improved to provide a 193 pixel map at eight imaging levels. Initial tests of this system have been successful. Improved methods for extracting bubble dimensions and rise velocities will be explored with the higher resolution system. Experiments are planned to study a variety of fluidization phenomena. Long term plans include further improvements in spatial resolution and the construction of an imaging system for a 30 cm diameter vessel.

ACKNOWLEDGEMENTS

The authors wish to acknowledge the dedicated assistance of members of the Instrumentation and Control Branch at METC including Dr. Nelson Smith, Ms. Jean Loudin, Mr. Carroll Utt, Mr. John Rotunda, Mr. John Trader, Mr. Keith Dodrill, and Ms. Kim Witherow.

NOTATION

D_c -- diameter of cylindrical section of slug determined from voidage peak height for 1/8-inch nylon sphere experiments, cm

D_s -- diameter of a sphere with a volume equal to the bubble, cm

$(D_s)_{i-1}$ -- diameter of bubble immediately ahead of bubble under consideration, cm

D_t -- tube diameter, cm

g -- acceleration of gravity, cm/sec^2

H_{max} -- maximum height of slugging bed, cm

H_{mf} -- height of bed at minimum fluidization, cm

L_s -- length of a slug calculated from its volume and D_c -cm

S_p -- spacing from base of bubble (peak voidage point) to base of proceeding bubble (its peak voidage point) times rise velocity of proceeding bubble, cm

U_a -- rise velocity of bubble or slug determined from time shift in average voidage plots, cm/sec

U_{mf} -- minimum fluidization velocity, cm/sec

U_o -- superficial gas velocity, cm/sec

V -- volume of slugs measured for 1/8-inch nylon sphere experiments, cm^3

LITERATURE CITED

1. Fasching, G.E., and N.S. Smith (1988), "High Resolution Capacitance Imaging System," NTIS Report No. DE88010277.

2. Davidson, J.F., and D. Harrison (1971), "Fluidization," Academic Press, London.

METAL CAPTURE DURING FLUIDIZED BED INCINERATION OF SOLID WASTES

T.C. Ho, J.M. Chen, S. Shukla and J.R. Hopper ■ Department of Chemical Engineering, Lamar University, Beaumont, TX 77710

The objective of this work was to study experimentally the characteristics of metal capture by bed sorbents during fluidized bed incineration of solid wastes contaminated with heavy metals. Experiments were carried out in a 7.62 cm laboratory scale fluidized bed of various sorbents, including limestone, bentonite, and aluminum oxide. Artificial test material contaminated with lead was prepared and incinerated in the bed with each sorbent under different operating conditions. An atomic absorption spectrometer was employed to measure the lead concentration in the original test material, the fresh sorbent and the incinerated bottom ash. The effectiveness of metal capture by each sorbent under specific operating conditions was evaluated based on the observed lead concentration. The results have indicated that the concept of metal capture by bed sorbent during fluidized bed incineration is highly promising.

The emission of toxic metals during the incineration of solid wastes containing metals presents a potential health hazard to human beings. The emitted metals, such as arsenic, barium, beryllium, chromium, cadmium, lead, mercury, nickel, and zinc, are often concentrated on particles with diameters of less than 1μm (Davidson et al., 1974). This significantly enhances the threat posed by the metals as air pollution control equipment is less efficient at capturing these small particles and particles of this size are most likely to be deposited in the lungs (see, e.g.,Oppelt, 1987).

One of the alternative control technologies for heavy metal emissions is the capture of metals during incineration. An incinerator with adequate residence time, optimum temperature and intimate contacting of solid sorbent and metallic substance, is required for the effective capture of the metals. Of available incineration systems, the fluidized bed incinerator appears to be best suited for the purpose.

Fluidized beds have been widely employed in the chemical industry to

*Department of Chemistry, Lamar University

carry out gas-solid reactions, such as calcination, gasification, and catalytic reaction (see, e.g., Kunii and Levenspiel, 1969). The inherent features that render a fluidized bed suitable for carrying out these reactions also make it an ideal choice for solid-waste incineration. The major features of the process include intensive mixing, effective gas-solid contact, high heat and mass transfer rates, high thermal stability and uniform temperature. These features enable the fluidized bed incinerator to operate smoothly and efficiently at low incineration temperatures without generating hot spots. The low incineration temperature suppresses the formation of NOx and reduces the possibility of ash melting. Although the research in the field of fluidized bed incineration has become more and more intense, relatively little work has been reported on metal emissions control using such a system.

It was the objective of this work to study experimentally the characteristics of metal capture by bed sorbents during fluidized bed incineration of solid wastes contaminated with heavy metals. Experiments were carried out in a 7.62 cm laboratory scale fluidized bed of various sorbents, including limestone, bentonite, and aluminum oxide.

Artificial test materials contaminated with lead were prepared and incinerated in the bed with each sorbent under different operating conditions. An atomic absorption spectrometer was employed to measure the lead concentration in the original test material, the fresh sorbent, and the incinerated bottom ash. The effectiveness of metal capture by each sorbent under specific operating conditions, was evaluated based on the observed lead concentrations.

THEORETICAL

Metal vapors (pure metals, oxides, chlorides, etc.) are generated through vaporization, decomposition and reaction in a high temperature combustion chamber following thermodynamic equilibrium and reaction kinetics laws (Barton et al., 1988). These generated metal vapors will condense in the combustion chamber whenever the partial pressure exceeds the vapor pressure. This will occur if the reactions involve the generation of new species which are relatively nonvolatile. This will also occur when the metal vapors move away from the burning waste to the relatively cooler surroundings. The drop in temperature will reduce the vapor pressure and cause condensation. Two mechanisms are associated with the condensation process, namely homogeneous nucleation and heterogeneous deposition (see, e.g., Friedlander, 1977; Lee, 1988).

Homogeneous Nucleation

When condensation takes place through the generation of new nuclei by the vapor itself, the process is known as homogeneous nucleation. Homogeneous nucleation may occur when the partial pressure of an inorganic vapor species exceeds a certain critical value (see, e.g., Friedlander, 1977; McNallan et al., 1981). The critical supersaturation that is necessary is dictated by the surface and volume free energies of the phase that may be nucleated. The incineration gases may become super-saturated simply as a result of rapid cooling of the gas or rapid formation of a new and relatively nonvolatile species.

Heterogeneous Deposition

When large concentrations of particles are present and the supersaturation is low, condensation takes place on the available surfaces without formation of new nuclei. This process is termed heterogeneous deposition (see, e.g., Friedlander, 1977). Particles in the incineration chamber or chamber walls may be sites for the heterogeneous condensation of the condensable species.

Metal Emissions

Heterogeneous and homogeneous condensations occur simultaneously at different rates. Homogeneous nucleation in incineration gases may explain the presence of the large number density of very fine metal particulates (micron and submicron size) that are found in the effluent gases of many incineration units. Particles in these size ranges are particularly troublesome because they are least likely to be captured completely in pollution abatement equipment and are most likely to be deposited in the lungs.

Heterogeneous deposition usually occurs on larger particles, which can be effectively collected or captured by the air pollution control devices. Therefore, to promote heterogeneous deposition would practically limit the formation of fine metal particulates and minimize their impact on the environment and on the health of the human population.

Concept of Metal Capture

The concept of metal capture is to promote heterogeneous deposition of metal vapors on sorbents whenever condensation occurs during incineration. The concept is especially promising when a fluidized bed combustor is employed, i.e., an effective contact between metal vapors and sorbents is provided. This work was designed to demonstrate the concept and to evaluate the effectiveness.

EXPERIMENTAL

The metal capture experiments were carried out in a 7.62 cm (3") fluidized bed. The experimental apparatus and procedure are described below.

Apparatus

A schematic diagram of the experimental apparatus is shown in Figure 1. It includes a fluidized bed preheater, a 7.62 cm (3") fluidized bed incinerator assembly, and a pressure and temperature monitoring and recording assembly.

The incinerator assembly was fabricated from stainless steel and consisted of a plenum section, a distributor, a combustor section and two cyclones. The height of the combustion section was 61 cm (24"). Associated with the combustion section was a waste feed tube at the top, an ash discharge tube at the bottom through the distributor and a side stream tube 14 cm (5.5") above the distributor. The distributor was a perforated plate with a 2.6% open area ratio. It was covered with fine inconel gauze to prevent the backflow of the bed material into the plenum chamber. The electric heaters controlled by six powerstat controllers, were applied to the outside surfaces of the columns. The incinerator was heavily insulated with high density Kao-Wool blankets.

The pressure and temperature monitoring and recording assembly consisted of pressure taps along the combustion and plenum sections, pressure transducers, a strip chart recorder, eight type-K thermocouples at different heights and an IBM PC/XT. The pressure signals were recorded using the strip chart recorder. The analog temperature signals were converted into digital signals through a data acquisition system and then stored and printed by the IBM PC/XT.

Experimental Procedure

For each experimental run, a designed amount of test material was charged into the bed under specific operating conditions. The dynamic temperature and pressure readings were monitored and recorded during incineration. The bottom and fly ash were collected after the incineration was completed. The collected ash was then sampled, acid extracted, and analyzed to determine its metal concentration employing an atomic absorption spectrometer (Perkin-Elmer Model 2100).

Artificial Test Material

The artificial test material used in this study was prepared from cylindrical wood pieces of 0.95 cm (3/8") in diameter * 0.64 cm (1/4") in length. The wood pieces were immersed in a lead nitrate solution for several days and were then dried to simulate lead contaminated wastes. The original lead content in the wood pieces was about 35 ppm (by weight) and the final lead content in the contaminated wood pieces ranged from 300 to 7000 ppm.

Fluidized Bed Sorbents

Three sorbents, serving as both the fluidized particles and the metal capture agents, were tested in the study. They were limestone, bentonite, and aluminum oxide. These sorbents were selected based mainly on one or more of the following considerations; they are inexpensive, have high temperature stability, are composed of porous structure, are readily available, and have chemical reactivity. Table 1 summarizes the size groups and the physical properties of each sorbent.

Specific Capture Capacity (ϕ)

The specific capture capacity is defined to be the amount of lead captured by a unit mass of bottom ash. It can be calculated as

$$\phi = Pb_b - \frac{Pb_o}{(1 - DR)} \quad (1)$$

where DR represents the average fractional loss of sorbent mass during incineration through calcination.

Percent Captured (ψ)

The percent captured is defined to be the percent of lead captured by fluidized bed sorbents over the total amount of lead charged. It can be calculated from the following equation:

$$\psi = \frac{\phi * W_b}{Pb_w * W_w} * 100\% \qquad (2)$$

RESULTS AND DISCUSSION

The section is divided into two sub-sections: general results and effects of operating conditions.

General Results

Tables 2 through 5 summarize typical experimental results on metal capture. Included in the tables are operating conditions as well as the percent lead captured by sorbents and the specific capture capacity of sorbents. In general, the results indicate that metal capture by fluidized bed sorbents is promising. The effectiveness, however, varies with sorbent species, sorbent size, sorbent amount, metal concentration in the waste, metal to sorbent ratio, air velocity, and temperature. The results also indicate the existence of some uncontrollable experimental factors.

Table 2 shows typical results of limestone experiments. The lead concentration in the test material for this set of experiments is relatively low at 337 ppm. The average DR for limestone is 0.41 at 900 $^\circ$C and 0.25 at 750 $^\circ$C. The results indicate that lead captured by limestone ranges from 43% to 89% with the specific capture capacity ranging from 17.1 to 37.6 µg/g of incinerated bottom ash. Better capture efficiencies are associated with lower temperatures, smaller particles, larger sorbent amount and higher air flow rates.

Tables 3 and 4 display typical results of calcinated limestone and bentonite experiments, respectively. The lead concentration in the test material for these experiments is relatively high at 4924 ppm. Note that the two sorbents were pre-calcinated in a furnace at 900 $^\circ$C for two hours before being tested and no further decomposition loss were detected during incineration. The results indicate that, with this relatively higher lead concentration in the waste, the percent of lead captured by sorbents is lower ranging from 23% to 70% compared to those in Table 2, i.e., 43% to 89%. The specific capture capacity, however, is higher for these experiments ranging from 76.7 to 593.7 compared to those in Table 2, i.e., from 17.1 to 37.6. The effectiveness of metal captured by the two sorbents is approximately equal. Better capture efficiencies are observed to be associated with smaller particles, which is consistent with that observed in Table 2. In addition, an increase in the amount of test material results in a better percent captured and higher specific capture capacity.

Table 5 summarizes typical results associated with the aluminum oxide. The test materials involved in these experiments have the same lead concentrations as those in Tables 3 and 4 at 4924 ppm. The results indicate that the capture efficiency for the aluminum oxide is in general very high above 90%. The high efficiency may either suggest that aluminum oxide is an excellent sorbent or it may be attributed to aluminum oxide's being the smallest sorbent tested.

One phenomenon, however, is worth pointing out associated with the aluminum oxide experiments. It was observed that a large fraction of the sorbent adhered to the chamber wall during incineration. The average amount of aluminum oxide collected immediately after the incineration was about 200 g, which accounted for only about 50% of the sorbent originally charged. The uncertainty as to the exact amount of aluminum oxide involved in the sorption process would inevitably have some effect on the reliable estimation of the capture efficiency.

Nonetheless, the results in Tables 2 through 5 clearly indicate that the concept of metal capture is practical. For example, the lead concentration of aluminum oxide is observed to increase from an initial value of 35.5 ppm before the incineration to a final value of 1408 ppm (Experiment Number 53) after the incineration, indicating a metal capture of up to 1372.5 ppm. Additional experiments, however, are needed to provide quantitative analysis.

Effects of Operating Conditions

An attempt was made to analyze the data and to evaluate trends regarding the effects of operating conditions on the effectiveness of metal capture. The results are discussed below.

Temperature Effect

Figure 2 plots the effect of temperature on capture efficiency for several experiments. It is observed that the efficiency is lower at a higher temperature. This is expected since at higher temperatures, the vapor pressure is higher and more metals escape as vapors, i.e., less condensation on sorbents.

Sorbent Size Effect

Smaller size sorbents have higher capture efficiency as indicated in Figure 3. This is due to the fact that smaller particles have greater surface areas, i.e., more surfaces to capture metal vapors.

Air Velocity Effect

Figure 4 plots the effect of air velocity on the capture efficiency for limestone. The results indicate that the efficiency is slightly higher at 40 cm/sec than that at 20 cm/sec. This is because that better mixing environment is generated at 40 cm/sec, providing better contact between metal vapors and sorbents.

Waste Amount Effect

As the amount of test material increases, the amount of metal capture increases as indicated in Figures 5 and 6. The change of slope toward better capture efficiency at higher amount charged as shown in Figures 5 and 6 appears to suggest that the sites of sorbent surface where metals have previously condensed may be more active in absorbing additional metal vapors. The phenomenon, however, requires additional experiments to confirm it.

Sorbent Species Effect

The current experimental data are insufficient for making a meaningful discussion as to the effect of sorbent species on the capture effectiveness. Although the highest capture efficiency was observed for aluminum oxide, the cause may be attributed to its being the smallest sorbent. Additional experiments are needed to provide better evaluations.

CONCLUSIONS

A study of metal capture by sorbents during fluidized bed incineration of metal contaminated solid waste has been carried out. The study investigates the effectiveness of metal capture by various sorbents under different operating conditions. Specifically, the study has resulted in the following conclusions:

1. The concept of metal capture is promising. The capture efficiency varies with sorbent species, sorbent size, sorbent amount, metal concentration in the waste, waste to sorbent ratio, air velocity and temperature.

2. The observed data indicate that better capture efficiencies are associated with smaller sorbent particles, larger sorbent amount, lower incineration temperature, better incinerator mixing, and higher amount of waste charged within the experimental ranges studied.

3. Additional experiments are needed to confirm current observations and to provide systematic data for developing models which are capable of analyzing the process quantitatively.

ACKNOWLEDGEMENT

The authors are grateful for the financial support of this work by the U. S. EPA through the Gulf Coast Hazardous Substance Research Center located at Lamar University (Grant No. 108LUB0114CF1).

NOTATION

$\bar{d}$ average particle diameter, mm

DR   average fractional calcination loss

Pb_o lead concentration in original sorbent, ppm

Pb_b lead concentration in bottom ash, ppm

Pb_w lead concentration in wood, ppm

T   temperature, °C

U   superficial velocity, cm/sec

W_b mass of collected bottom ash, g

W_s mass of sorbent, g

W_w mass of wood, g

ψ percent captured

ϕ specific capture capacity, μg/g of bottom ash

ρ density, g/cm³

LITERATURE CITED

Barton, R. G., P. M. Maly, W. D. Clark, and W. R. Seeker, "Prediction of The Fate of Toxic Metals in Waste Incinerators," ASME 13th National Waste Processing Conference, 379-386 (1988).

Davidson, R. L., David F. S. Natusch, John R. Wallace, Charles A. Evans, "Trace Elements in Fly Ash Dependence of Concentration on Particle Size," Environmental Science & Technology, 8, No.13, 1107-1113 (1974).

Friedlander, S. K., Smoke, Dust and Haze Fundamentals of Aerosol Behavior, John Wiley & Sons, Inc., 1977.

Kunii, D., and O. Levenspiel, Fluidization Engineering, Johe Wiley and Sons, Inc., New York, Ny, 1969.

Lee, C. C., "A Model Analysis of Metal Partitioning," JAPCA, 38, No.7, 941-945 (1988).

McNallan, M. J., G. J. Yurek, and J. F. Elliott, "The Formation of Inorganic Particulates by Homogeneous Nucleation in Gases Produced by the Combustion of Coal," Combustion and Flame, 42, 45-60 (1987).

Oppelt, E. T., "Incineration of Hazardous Waste - A Critical Review," JAPCA, 37, No.5, 558-586 (1987).

Table 1. Properties of Fluidized Bed Particles

Particle (Mesh No)	$\bar{d}$	ρ	U_{mf}		
			25°C	750°C	900°C
Limestone (#20-#30)	0.7	2.48	32.0	16.6	15.4
Limestone (#30-#40)	0.5	2.48	18.4	8.6	7.9
Calcinated Limestone (#20-#30)	0.7	1.50	20.8	10.1	9.4
Bentonite (#20-#40)	0.6	2.20	22.6	10.9	10.1
Bentonite (#35-50)	0.4	2.20	10.9	4.9	4.5
Aluminum Oxide (#70-#80)	0.3	4.00	11.2	5.0	4.6

Table 2. Typical Experimental Data for Limestone
(Pb_o=51.3 ppm, Pb_w=337 ppm, DR=0.41 at 900 °C, DR=0.25 at 750 °C)

Exp. No.	T (°C)	U (cm/s)	$\bar{d}$ (mm)	W_s (g)	W_b (g)	W_w (g)	Pb_b (ppm)	ϕ (μg/g)	ψ (%)
11	900	20.2	0.7	600.4	321.3	51.0	110	23.0	43
15	900	19.1	0.7	600.4	321.2	30.6	104	17.1	53
12	900	38.6	0.7	600.4	313.1	50.9	120	33.1	60
8	750	21.2	0.5	600.5	396.4	50.8	106	37.6	87
16	750	20.2	0.5	600.5	396.4	30.2	90	21.6	84
18	900	20.6	0.5	602.9	322.6	30.4	106	19.1	60
17	750	40.6	0.5	600.0	355.1	30.3	94	25.6	89
19	900	40.7	0.5	600.2	313.0	30.1	110	23.0	71
22	750	39.4	0.5	550.6	325.9	30.1	91	22.6	73
27	750	40.5	0.7	554.1	327.9	30.1	90	21.6	70

Table 4. Typical Experimental Data for Bentonite
(Pb_o=43.3 ppm, Pb_w=4924 ppm, DR=0)

Exp. No.	T (°C)	U (cm/s)	$\bar{d}$ (mm)	W_s (g)	W_b^* (g)	W_w (g)	Pb_b (ppm)	ϕ (μg/g)	ψ (%)
47	750	39.5	0.6	301.2	301.2	20.2	120	76.7	23
48	750	40.2	0.6	304.7	304.7	40.5	321	277.7	42
54	750	39.7	0.4	270.9	270.9	40.1	547	503.7	69
49	750	20.2	0.4	305.5	305.5	60.1	637	593.7	61

*W_b fluctuates around W_s at ± 5% difference.

Table 3. Typical Experimental Data for Calcinated Limestone
(Pb_o=69.4 ppm, Pb_w=4924 ppm, DR=0)

Exp. No.	T (°C)	U (cm/s)	$\bar{d}$ (mm)	W_s (g)	W_b^* (g)	W_w (g)	Pb_b (ppm)	ϕ (μg/g)	ψ (%)
43	750	40.0	0.7	300.1	300.1	20.4	157	87.6	26
44	750	39.1	0.7	303.4	303.4	40.1	259	189.6	29
45	750	38.9	0.7	302.6	302.6	61.5	465	395.6	40

*W_b fluctuates around W_s at ± 5% difference.

Table 5. Typical Experimental Data for Aluminum Oxide
(Pb_o=35.5 ppm, Pb_w=4924 ppm, DR=0)

Exp. No.	T (°C)	U (cm/s)	$\bar{d}$ (mm)	W_s (g)	W_b^* (g)	W_w (g)	Pb_b (ppm)	ϕ (μg/g)	ψ (%)
51	750	21.1	0.3	399.5	200.0	20.2	517	481.5	97
52	750	21.2	0.3	399.7	200.0	40.0	918	882.5	90
53	750	21.2	0.3	401.0	200.0	60.3	1408	1372.5	93

*W_b is the average observed value.

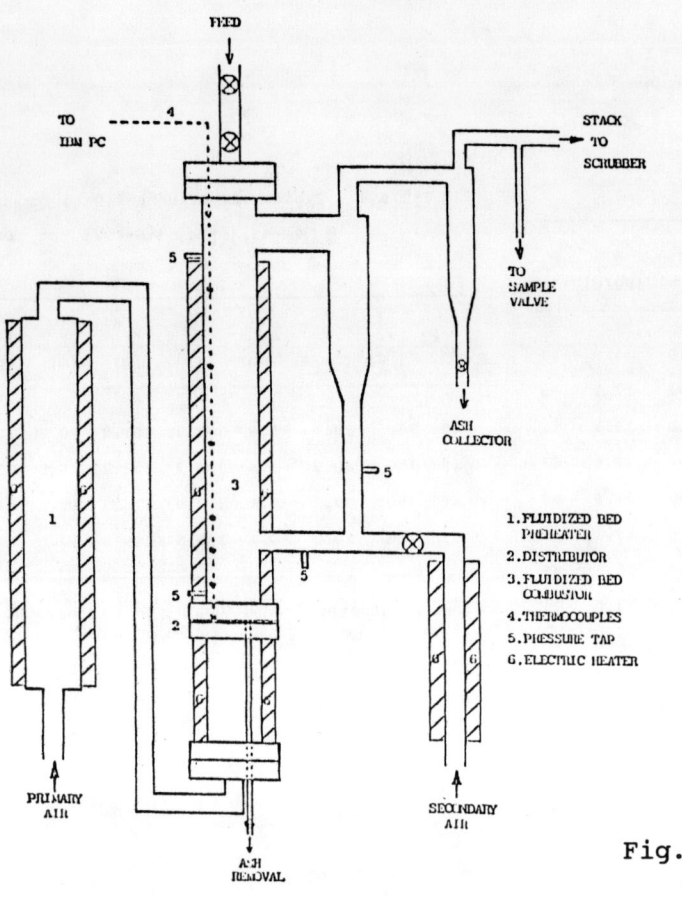

Fig. 1. Incinerator set-up.

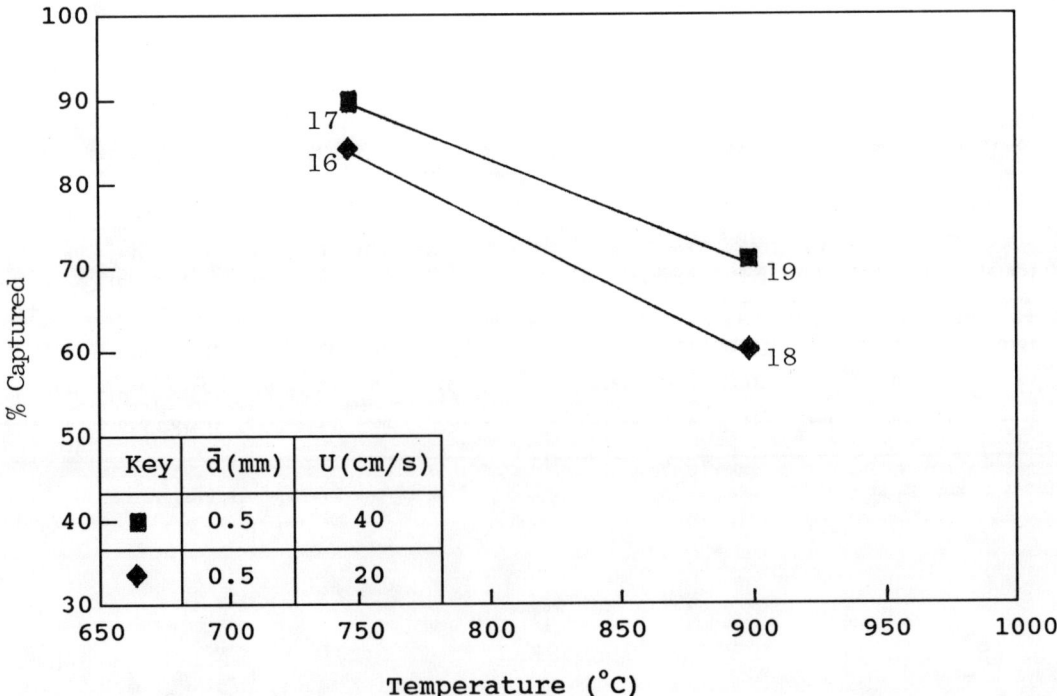

Fig. 2. Effect of temperature on the percent captured.

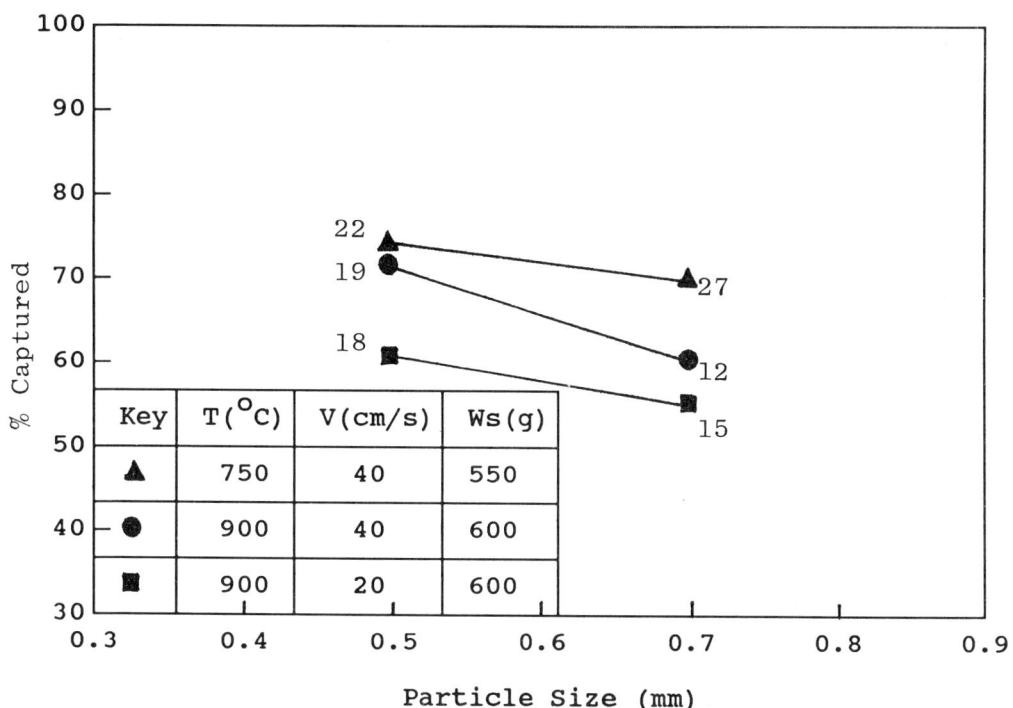

Fig. 3. Effect of particle size on the percent captured.

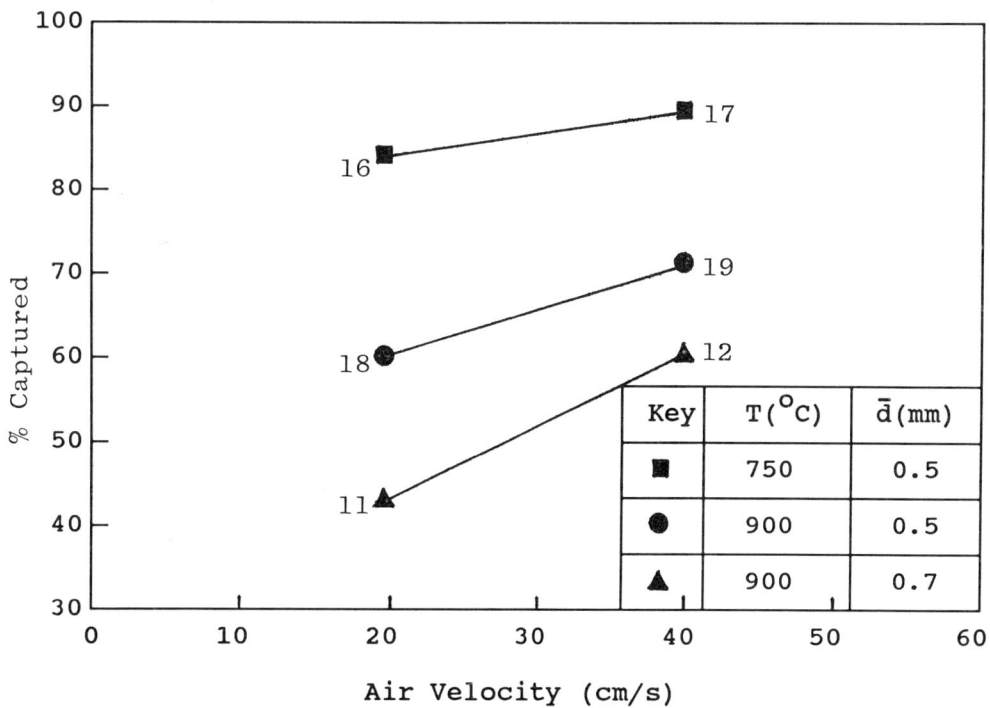

Fig. 4. Effect of air flow rate on the percent captured.

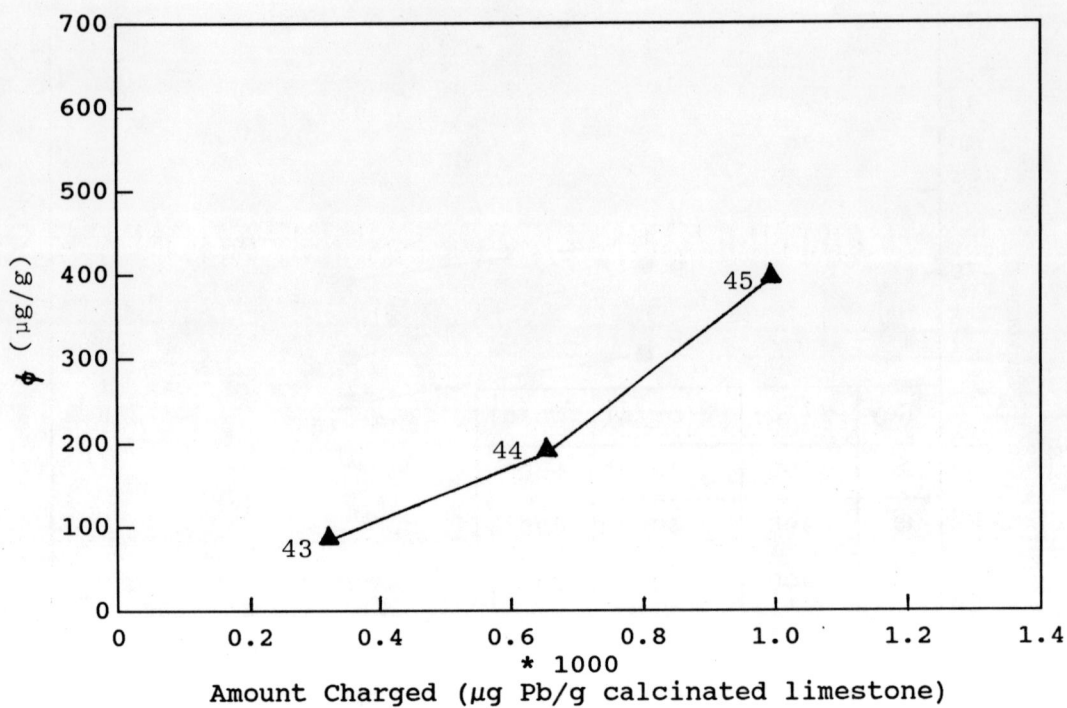

Fig. 5. Effect of the amount lead charged on the specific capture capacity for calcinated limestone.

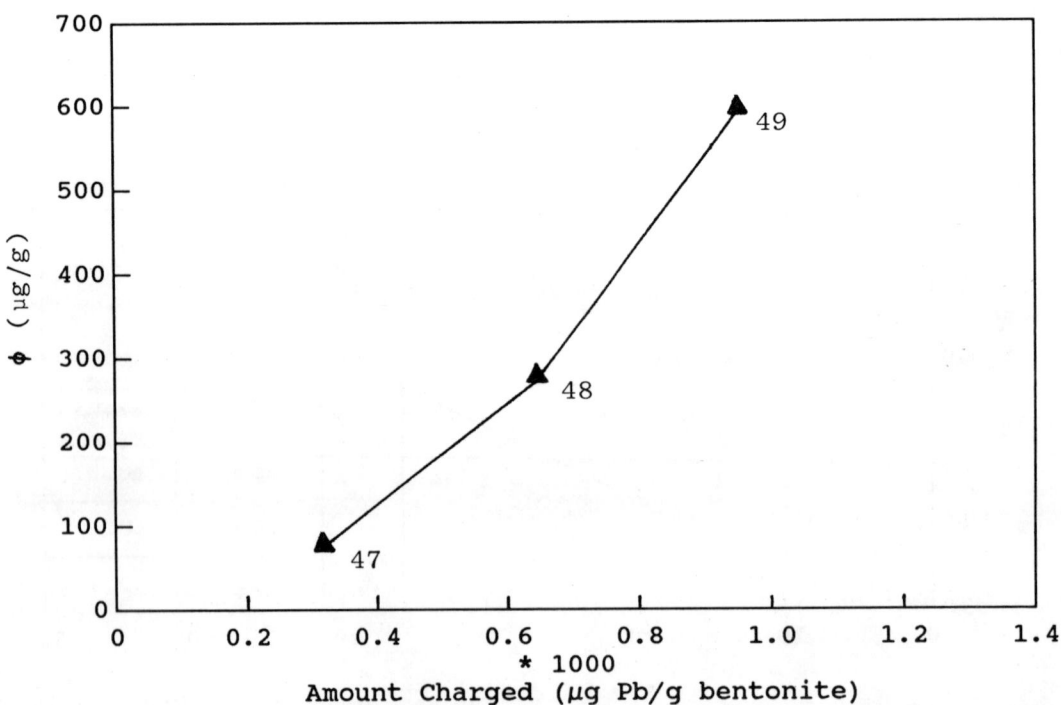

Fig. 6. Effect of the amount lead charged on the specific capture capacity for bentonite.

CONTINUOUS DEPRESSURIZATION OF SOLIDS USING A RESTRICTED PIPE DISCHARGE SYSTEM

T.M. Knowlton and J.G. Findlay ■ Institute of Gas Technology, Chicago, IL
I. Chan ■ M.W. Kellogg Co., Houston, TX

A novel Restricted Pipe Discharge System (RPDS) was used to depressure 224- to 621-micron oil shale from 600 psig to atmospheric pressure. The RPDS was found to smoothly and reliably depressurize the solids. It has an advantage over a lockhopper as a depressurization device, because solids discharge is continuous, valve maintenance costs are lower, solids can be transferred and depressurized simultaneously, and capital costs are less.

INTRODUCTION

For many years the lockhopper has been the traditional device used to transfer dry solids from a high-pressure vessel to atmospheric pressure. However, lockhoppers have several disadvantages: 1) the transfer of solids is discontinuous, and 2) capital and maintenance costs for lock-hopper valves are generally high (especially if the solids are abrasive and the system is hot). In a restricted pipe discharge system (RPDS), solids are simply discharged in moving packed-bed flow from system pressure to atmospheric pressure through a pipe restricted at the outlet. The pressure drop in the pipe is produced by the gas flowing faster than the solids. Systems using this type of solids flow were previously studied by Berg (1, 2, 3, 4), and more recently by Knowlton, et al. (5). The potential advantages of the RPDS over the lockhopper are lower capital cost, continuous solids discharge, much lower valve maintenance costs (cycling on-off valves are not required), and lower process gas loss.

EQUIPMENT DESCRIPTION AND OPERATION

Three different RPDS configurations were tested in this investigation. A flow sheet of the first RPDS configuration tested in the program is shown in Figure 1. This configuration was called the "vertical downward" configuration. The dimensions of this test configuration are given in Figure 2. The test unit basically consisted of a pressure vessel, the depressurization piping, a disengaging vessel, the solids control valve, and a solids recycle system. The pressure vessel was 9.5 feet high and 36 inches in diameter, and was capable of operating at pressures up to 650 psig. The pressure vessel was equipped with a 60-degree-angle internal cone to facilitate solids removal at the bottom. The gas-solid disengaging vessel was 3.5 feet high and 18 inches in diameter.

In operation, solids flowed out of the pressure vessel through a section of 2-inch-diameter pipe, and then a section of 3-inch-diameter pipe, before discharging into the disengaging vessel. The two-diameter depressurization pipe was selected to determine the effect of pipe diameter on solids flow and pressure drop in the system. A packed bed of solids was maintained in the disengaging vessel. This packed bed acted as a solids flow restriction at the end of the pipe. A 3-inch-diameter slide valve located below the disengaging vessel, was used to control the solids flow rate in the system.

The solids flowed through the slide valve into a 2-inch-diameter pneumatic conveying line which carried the solids up into a 4-foot-diameter solids receiver vessel. The receiver vessel was used to store the solids discharged from the disengaging vessel during a test, and was maintained at atmospheric pressure. Following a test, the feed vessel was depressurized, and the 3-inch-diameter ball valve located between the receiver and the pressure vessel was opened to allow the solids to flow into the pressure vessel.

Solids flow rates in the system were determined by diverting and collecting the solids in a container over a measured time interval, and then weighing the solids. This was accomplished by using the two ball valves located below the 3-inch-diameter slide valve.

The gas flowing with the solids through the slide valve was essentially flowing at the same velocity as the solids, and was negligible compared to the gas flow out of the disengaging vessel. Assuming a voidage of 0.5, the amount of gas flow through the slide valve at a solids flow rate of 6000 lb/h would be

$$6000 \frac{lb}{h} \times \frac{1}{74.5} \frac{ft^3}{lb} \times \frac{1}{60} \frac{h}{min} \times 0.5$$

$$= 0.67 \text{ scfm}$$

This is much less than the gas flow measured at any test pressure at the same solids flow rate (Figure 9). Nitrogen gas was used as the motive gas in the system. It was added to the pressure vessel near the top of the vessel. A backpressure regulator was used to maintain a constant pressure in the pressure vessel. Gas from the disengaging vessel was discharged from the system at atmospheric pressure and was metered with a rotameter.

Pressure taps were located along the entire flow path. The taps were connected to differential-pressure transmitters to obtain pressure drops between various points in the system. These differential pressures were recorded in all tests.

A second piping configuration was also investigated in the RPDS testing. Its configuration and dimensions are shown in Figures 3 and 4, respectively. This was a two-stage, vertical downward configuration in which system gas was removed not only at near atmospheric pressure, but also simultaneously at an intermediate pressure. This configuration was called the "two-stage" configuration.

A third piping configuration that was also investigated in the RPDS test program is shown in Figure 5. This configuration was called the "vertical transfer" configuration. The dimensions of this test unit are given in Figure 6. In this configuration, solids first flowed downward out of the pressure vessel into a short, vertical section of 2-inch-diameter pipe, and then into a horizontal section of 2-inch-diameter pipe. The solids then flowed upward through a long vertical section of 2-inch-diameter pipe. Following this section, the solids passed through a horizontal section of 2-inch-diameter pipe, and then through a vertical downflow 3-inch-diameter pipe section back into the disengaging vessel. The 3-inch-diameter slide valve located below the disengaging vessel controlled the solids flow rate in the system.

EXPERIMENTAL PROCEDURE

In a typical test, the shale test material was first loaded into the 3-foot-diameter pressure vessel. The 3-inch-diameter slide valve and the 3-inch-diameter ball valve were closed. The 2-inch-diameter ball valve below the pressure vessel was then opened, and the unit was pressurized. During pressurization, the solids flowed through the pressure-reduction piping and into the gas-solid disengaging vessel. The flow of solids ceased when a packed bed of solids was established in the disengaging vessel. After reaching the desired system pressure, the steady-state pressure drops between various sections of the pressure-reduction piping, the gas flow out of the disengaging vessel, and other system parameters were recorded with no solids flow in the unit. The air flow in the 2-inch-diameter pneumatic conveying line was set at a conveying line velocity of approximately 30 to 35 ft/s. Solids flow was then started by opening the 3-inch-diameter slide valve. The opening of the slide valve was adjusted until the desired solids flow rate was achieved. Measurement of the solids flow rate was accomplished by diverting and collecting the solids over a short period of time. At each solids flow rate, pressure drops, exit gas flows out the disengaging vessel, and other system parameters were recorded. At the end of the test, the system was depressurized.

The operating procedure was basically the same for the "vertical downward" and the "vertical transfer" configurations. For the two-stage system, gas was discharged from the Stage 1 and Stage 2 disengaging vessels simultaneously. The pressure in Stage 1 was set at the desired constant value by setting the backpressure regulator in the gas outline line from the Stage 1 pressure vessel. All system pressure drops, and the gas flows from the Stage 1 and the Stage 2 disengaging vessels were recorded.

MATERIALS

Three different sizes of shale material were used in this investigation. The size range and the physical characteristics of each of these materials are given in Table 1. The minimum and complete fluidization velocities were experimentally determined for each of these solids in a 2-inch-diameter Plexiglas fluidization column. The values of these velocities are given in Table 1.

RESULTS AND DISCUSSION

Solids Flow Restriction

Two different types of solids flow restrictions were tested using the "vertical downward" RPDS configuration. The purpose of the test was to determine the best type of solids flow restriction to use inside the disengaging vessel. Tests were conducted at system pressures of 150 and 300 psig, and at nominal solids flow rates of 1500, 3000, and 6000 lb/h using -20+40 mesh shale.

A drawing of the first solids flow restriction tested is shown in Figure 7. It consisted of four gas-solid separation cylinders inserted into the disengaging vessel. Each cylinder was made of a continuous wire screen which allowed gas to pass through it, but not solids. The opening of the screen was approximately 51 microns wide. The gas-solid separation cylinders were spaced at 90 degrees around the vessel and concentric to the solids feed line into the vessel. During operation, the entire vessel was almost completely filled with the test solids.

Several tests were conducted with this configuration. In all of the tests, the system operated very smoothly. The pressure reading inside the disengaging vessel was steady in all tests. The solids flow in the system was stable and easily controlled.

A second type of solids flow restriction in the disengaging vessel was also tested. The second flow restriction, shown in Figure 8, was simply a packed bed of solids maintained in the bottom of the disengaging vessel. The advantage of this type of restriction is that it is extremely simple, just a bed of solids in a small vessel. Several tests were then conducted with this configuration using the same solids material that was used with the first configuration. Results of the tests showed that the packed-bed restriction was able to perform satisfactorily only at low system pressure (150 psig). At a pressure of 300 psig and a high solids flow rate (6000 lb/h), the gas flow (40.57 scfm) out of the disengaging vessel was large enough to cause local fluidization of the solids near the end of the 3-inch-diameter RPDS pipe. When this occurred the restriction to solids flow was destroyed, resulting in uncontrollable flow of the solids. This flow restriction was, therefore, considered to be unacceptable.

Effects of System Pressure and Solids Flow Rate

The effects of system pressure and solids flow rate on the operation of the RPDS were determined by testing at three different nominal pressures (150, 300, and 600 psig) and three different nominal solids flow rates (1500, 3000, and 6000 lb/h) using the "vertical downward" RPDS configuration. In each test, the gas flow rate out of the disengaging vessel, the pressure profile, and the pressure-drop-per-unit-length profile over the length of the pressure reduction pipe were determined.

Figure 9 shows the effect of system pressure on the total gas flow from the disengaging vessel for the three different solids flow rates tested. This figure shows that the amount of gas flowing out of the disengaging vessel increased with increasing system pressure and was found to be almost linearly proportional to system pressure. This result was expected, because in order to generate the additional pressure drop required at high pressures, gas flow in the RPDS pipe had to increase.

Increasing the solids flow rate from 1500 lb/h to 6000 lb/h was found to increase the gas flow from the disengaging vessel only slightly. This was because the gas velocity was much greater than the solid velocity inside the depressurization pipe. A typical gas velocity near the end of the 3-inch-diameter depressurization pipe at 1500 lb/h was calculated to be 14 ft/s at a system pressure of 300 psig. The velocity of the solids was 0.11 ft/s. Therefore, even when the solids velocity was increased fourfold to 0.44 ft/s (equivalent to a solids flow rate of 6000 lb/h), the gas velocity would only need to increase to 14.33 ft/s (14 ft/s + [0.44 ft/s - 0.11 ft/s]) in order to maintain the same gas-solids relative velocity and generate the same pressure drop. This represented only a 2.4% increase in the gas flow rate.

Pressure profiles, for the tests conducted at three different pressures and at a solids flow rate of 3000 lb/h, are shown in Figure 10 for the "vertical downward" RPDS configuration. The ΔP/L profiles at the different pressures used in these same tests are shown in Figure 11. These profiles were typical of those obtained in all of the tests conducted with the "vertical downward" configuration.

The data indicated that the pressure-drop-per-unit-length (ΔP/L) in the RPDS pressure reduction pipe was generally higher in the 2-inch pipe than in the 3-inch pipe. This is because both the relative gas-solids velocity and the gas density in the 2-inch pipe were higher than in the 3-inch pipe. The relationship between pressure-drop-per-unit-length, relative gas-solids velocity (V_r), and gas density (ρ_g) for gas flow through a packed bed of solids can be predicted by the Ergun Equation:

$$\frac{\Delta P}{L} = \frac{150 \, \mu \, (1-\epsilon)^2 \, V_r}{g_c \, (\phi \, D_p \, \epsilon)} + \frac{1.75 \, (1-\epsilon) \, \rho_g \, V_r^2}{g_c \, (\phi \, D_p \, \epsilon)} \quad (1)$$

where

$$V_r = V_g - V_s \quad (2)$$

Figure 11 also shows that pressure-drop-per-unit-length decreased dramatically as the diameter of the RPDS pipe was expanded from 2 inches to 3 inches. This was mainly due to the large drop in the gas velocity in the RPDS depressurization pipe as the area increased. Pressure-drop-per-unit-length was found to increase along the length of both the 2-inch and the 3-inch RPDS pipe. This result was expected because as the pressure decreases along the pipe, the gas expands, increasing the absolute gas velocity and also the relative gas-solids velocity.

Effect of Particle Size

Tests to determine the effect of particle size on the operation of the RPDS were also conducted with the "vertical downward" configuration. Tests with all three materials were made at nominal system pressures of 150, 300, and 600 psig.

A plot of the gas flow rate from the disengaging vessel versus particle size is shown in Figure 12 for the three pressures tested at a constant solids flow rate of 3000 lb/h. This figure shows that the amount of gas flow out of the disengaging vessel increases with increasing particle size. This is in agreement with the trend predicted by applying the Ergun Equation to the RPDS depressurization pipe. During the tests, it was also observed that the unit was noisiest when operating with the smallest material (224 microns) at the highest system pressure of 600 psig. This was probably due to the fact that solids flow in the RPDS depressurization pipe was in stick-slip flow at these conditions.

The ΔP/L distribution along the RPDS pipe is presented in Figure 13 for the three different shale particle sizes tested. Figure 13 indicates that the ΔP/L profiles did not change significantly as the particle size was changed.

Two-Stage Configuration

The RPDS was also modified to investigate the operability of a two-stage system (Figure 3). One of the potential benefits of the two-stage system is that it can be used to bleed off a portion of the gas in the depressurization line at a higher pressure than atmospheric. This high-pressure gas can then be used for other purposes, if desired.

Three tests were conducted with this configuration using the -20+40 mesh shale material. These tests were carried out at ambient temperature and at a system pressure of 300 psig. Three different solids flow rates (1500, 3000, and 6000 lb/h) were tested. In all tests, the Stage 1 disengaging vessel was maintained at a pressure of 80 psig, and the Stage 2 disengaging vessel at a pressure just slightly above atmospheric.

The results of the tests showed that the two-stage system was able to operate as smoothly as the single-stage system. Analysis of the data also showed that the total gas flow out of the two disengaging vessels was not much different than the total gas flow obtained earlier with the single-stage system, i.e., 42.4, 43.9, and 41.9 scfm for the two-stage configuration versus 43.0, 41.5, and 44.6 scfm for the single-stage configuration at solids flow rates of 1500, 3000, and 6000 lb/h, respectively.

A comparison of the pressure drop profiles for the two-stage and single-stage configurations is shown in Figure 14. This figure shows that the shape of the curves was similar for both systems. The ΔP/L profile for the two-stage system was found to be higher in the 2-inch section of the pressure-reduction pipe and lower in the 3-inch section of the pressure-reduction pipe, than the corresponding sections in the single-stage system. This is because the Stage 1 disengaging vessel in the two-stage system was maintained at a pressure of 80 psig, which was lower than the natural pressure (approximately 140 psig) developed at that location when operating with the single-stage configuration. Therefore, in the 2-inch section of the RPDS pipe and over the same pipe length, more pressure was dissipated in the two-stage system (300 to 80 psig) than in the single-stage system (300 to 140 psig). Conversely, in the 3-inch section of the RPDS pipe, less pressure was dissipated in the two-stage system (80 psig to near atmospheric pressure) than in the single-stage system (140 psig to near atmospheric pressure).

Vertical Transfer Configuration

The RPDS unit was also modified to determine the effect of solids flowing in vertical upflow on the operation of the RPDS (Figure 5). Six tests were conducted with this configuration using the -20+40 mesh shale material at nominal system pressures of 150 and 300 psig. Three different nominal solids flow rates (1500, 3000, and 6000 lb/h) were tested at each pressure.

During tests with the vertical transfer configuration, it was observed that the unit became noisier as the solids flow rate was increased. Pressure drop readings in the lower section of the vertical transfer pipe were found to decrease as the solids flow rate was increased, indicating that the solids were no longer in a moving packed-bed or stick-slip type of flow, but were probably travelling in slug flow.

The results of these tests are shown in Figures 15 and 16. In these figures, the pressure-drop profiles are plotted as a function of solids flow rate for each system pressure tested. In general, for the same transfer pipe diameter, the ΔP/L essentially increased continuously over the length of the pressure reduction line. However, as the solids flow rate was increased, the pressure-drop profile also began to shift. At the lowest solids flow rate (1500 lb/hr), the ΔP/L increased relatively evenly over the entire length of the 2-inch section of the RPDS pipe, indicating a moving packed-bed type of solids flow in the pipe. However, the ΔP/L distribution changed significantly when the solids flow rate was increased to 3000 lb/hr. It was found that the ΔP/L readings decreased in the lower part of the vertical transfer line as the solids flow rate was increased. The ΔP/L typically dropped to a reading of 0.5 psi/ft, which is slightly higher than the value (0.4 psi/ft) measured at minimum fluidization for these solids. Solids flow in that portion of the vertical transfer pipe was, therefore, considered to be in a slugging type of flow. It was observed that the length of the slugging portion of the line increased with increasing solids flow rate. Figures 15 and 16 also show that increasing the system pressure reduced the slugging in the vertical transfer line.

A plot of the amount of gas flow from the disengaging vessel versus system pressure for the vertical transfer configuration is shown in Figure 17 for the three different solids flow rates tested. The results show that, at all solids flow rates, increasing system pressure increased the gas flow from the RPDS pipe. This result was expected because similar results were observed when testing with the "vertical downward" RPDS configuration. However, with the "vertical transfer" configuration, it was found that increasing the solids flow rate also significantly increased the gas flow in the

RPDS pipe. This was not observed with the "vertical downward" configuration.

The primary reason for this was that the "vertical transfer" unit was much longer than the "vertical downward" unit (i.e., 49 feet versus 22.5 feet). Because it was longer, less gas was required to dissipate the same pressure. Therefore, the solids velocity in the pipe was a greater percentage of the gas velocity in the pipe. Therefore, when the solids flow rate was changed, the gas flow from the disengaging vessel had to increase more than in the case of the shorter "vertical downward" configuration.

CONCLUSIONS

1. The RPDS was found to operate very smoothly and reliably in removing solids from high-pressure environments.

2. The best type of solids flow restriction inside the disengaging vessel consisted of heavy-duty, gas-solid separation cylinders which allowed gas to pass through them, but not solids.

3. With the "vertical downward" RPDS configuration, it was found that -

    a) The amount of gas flowing from the disengaging vessel increased linearly with system pressure.

    b) For the length of the RPDS pipe tested, increasing the solids flow rate increased the gas flow from the disengaging vessel only slightly, and did not significantly affect the pressure-drop distribution.

    c) The $\Delta P/L$ increased along the length of the RPDS pipe due to gas expansion in the line.

    d) The amount of gas flowing out of the disengaging vessel increased with increasing particle size. This was in agreement with the trend predicted by the Ergun Equation.

4. The two-stage RPDS was able to operate as smoothly as the single stage. The total gas flow from the two-stage system was essentially the same as the gas flow rate measured from the single-stage system.

5. In the "vertical transfer" RPDS configuration, it was observed that -

    a) Increasing the solids flow rate significantly increased the gas flow from the system. This was not observed with the "vertical downward" configuration.

    b) Pressure drop readings were found to decrease in the lower part of the vertical transfer pipe as the solids flow rate was increased. Solids flow in that portion of the vertical transfer line was considered to be in slugging flow. The length of the slugging portion of the vertical transfer line increased with increasing solids flow rate and decreasing system pressure.

ACKNOWLEDGEMENT

The authors would like to express their appreciation to the U.S. Department of Energy which funded the work. The work was performed under Contract No. DE-AC21-87MC11089.

NOMENCLATURE

d_{pi} = particle diameter for ith discrete size fraction, ft
D_p = average particle diameter, ft
g_c = gravitational acceleration, 32.2 ft/s^2
U_{cf} = complete fluidization velocity, ft/s
U_{mf} = minimum fluidization velocity, ft/s
V_g = interstitial gas velocity, ft/s
V_r = relative gas-solids velocity, ft/s
V_s = solid velocity, ft/s
X_i = weight fraction of particles of size d_{pi}, dimensionless
$\Delta P/L$ = pressure-drop-per-unit-length, psi/ft
μ = gas viscosity, lb/s-ft
ϵ = voidage, dimensionless
ϕ = particle sphericity, dimensionless
ρ_g = gas density, lb/ft^3

LITERATURE CITED

1. Berg, C.H.O. (Assigned to Union Oil (Company of California), U.S. Patent 2,684,868 (1954).
2. Berg, C.H.O. (Assigned to Union Oil Company of California), U.S. Patent 2,684,870 (1954).
3. Berg, C.H.O. (Assigned to Union Oil Company of California), U.S. Patent 2,684,872 (1954).

4. Berg, C.H.O. (Assigned to Union Oil Company of California), U.S. Patent 2,684,873 (1954).
5. Knowlton, T., Findlay, J., Sishtla, C. and Chan, I., "Solids Pressure Reduction Without Lockhoppers: The Restricted Pipe Discharge System". Paper presented at the American Institute of Chemical Engineers National Meeting, Miami, Florida, November 2-7, 1986.

Table 1. PHYSICAL CHARACTERISTICS OF SOLIDS TESTED

Screen Size U.S.S. Mesh	Microns	Weight % Retained on Stated Mesh Size		
		-14+30 Shale	-20+40 Shale	-40+80 Shale
10	2000	--	--	--
12	1680	0.40	--	--
14	1410	0.56	--	--
18	1000	2.71	--	--
20	841	9.74	0.22	--
30	595	53.64	18.71	--
35	500	--	45.89	2.92
40	420	26.32	5.59	0.15
50	297	--	24.77	18.61
60	250	6.28	2.86	18.38
80	177	--	1.11	43.53
100	149	0.15	0.86	--
120	125	--	--	15.28
140	105	0.21	--	--
170	88	--	--	0.50
270	53	--	--	0.18
400	37	--	--	0.46
Total		100.00	100.00	100.00
Average Particle Size, D_p^*, μm		621	471	224
U_{mf}, ft/s		1.13	0.69	0.21
U_{cf}, ft/s		1.34	1.19	0.62

Particle Density = 128.6 lb/ft^3
Bulk Density = 74.5 lb/ft^3

* $D_p = \dfrac{1}{\sum X_i/dp_i}$

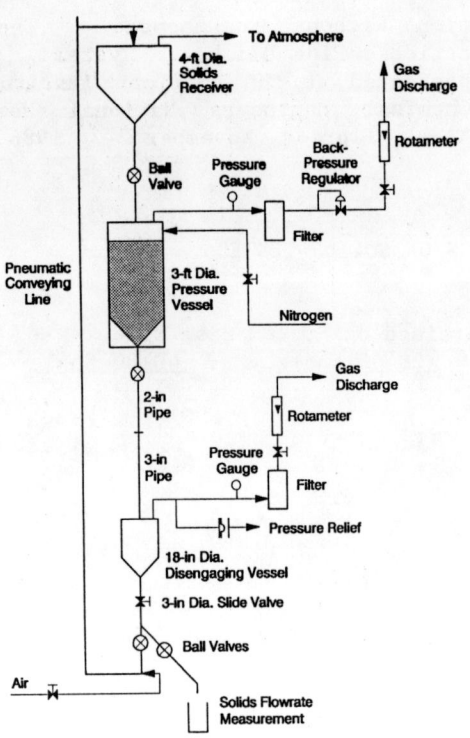

Figure 1. Schematic drawing of the vertical downward RPDS test configuration.

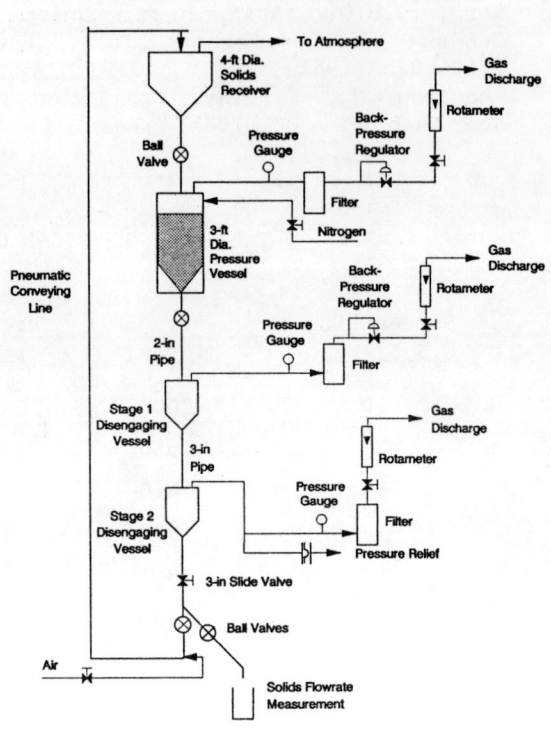

Figure 3. Schematic drawing of the two-stage RPDS test configuration.

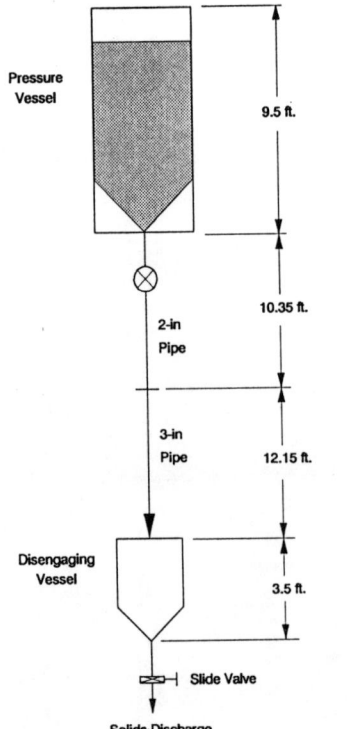

Figure 2. Vertical downward RPDS unit dimensions.

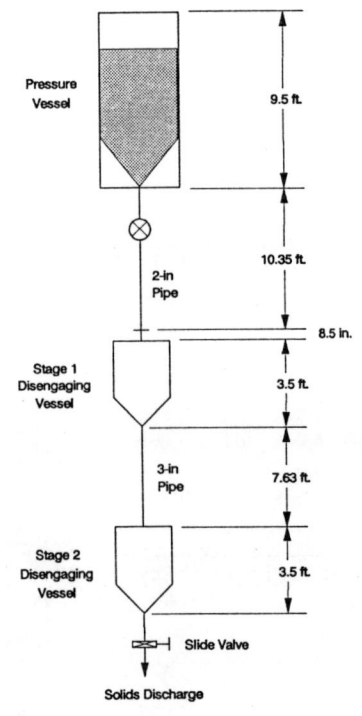

Figure 4. Two-stage RPDS unit dimensions.

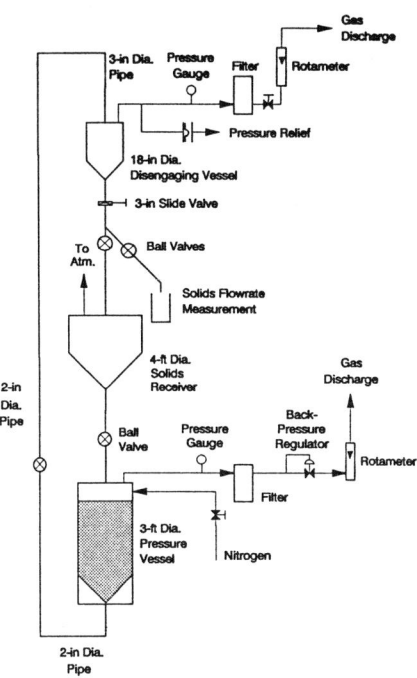

Figure 5. Schematic drawing of the vertical transfer test configuration.

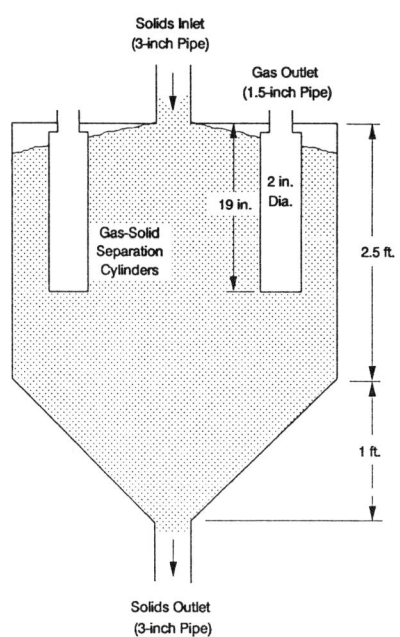

Figure 7. Schematic drawing of disengaging vessel with screen barrier restriction.

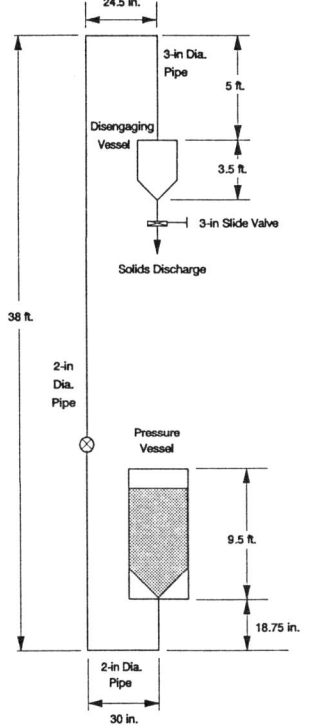

Figure 6. Vertical transfer RPDS unit dimensions.

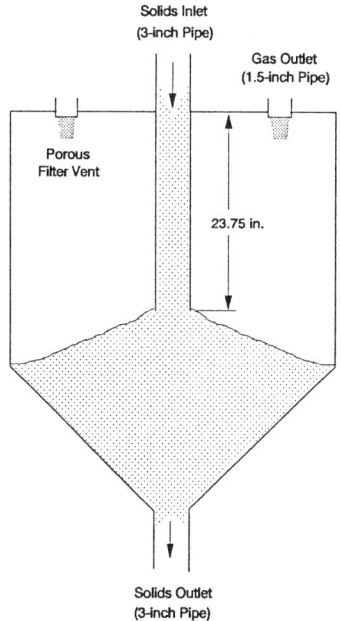

Figure 8. Schematic drawing of disengaging vessel without screen barrier restriction.

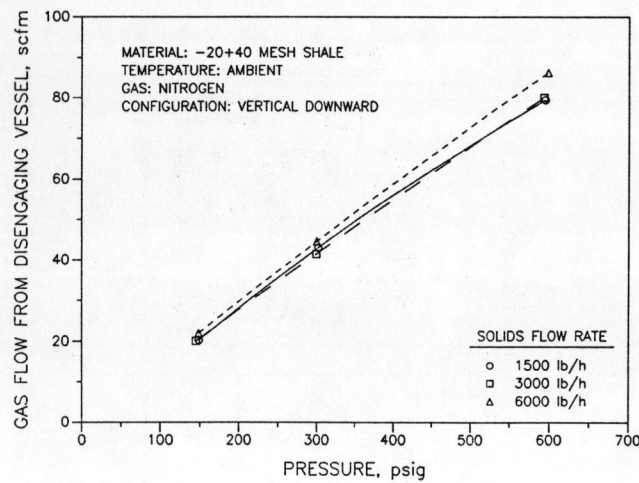

Figure 9. Gas flow from disengaging vessel as a function of pressure.

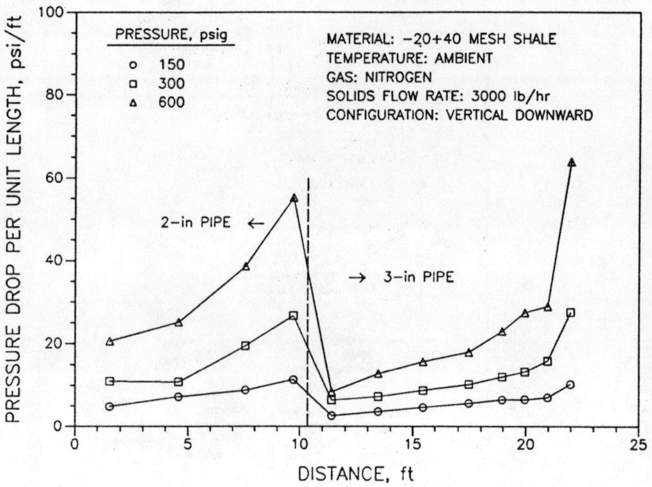

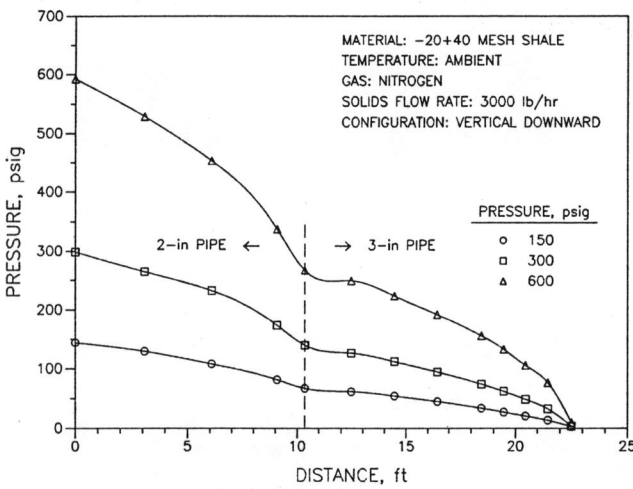

Figure 10. Pressure profiles along the depressurization pipe as a function of system pressure.

Figure 11. Pressure-drop profiles along the depressurization pipe as a function of system pressure.

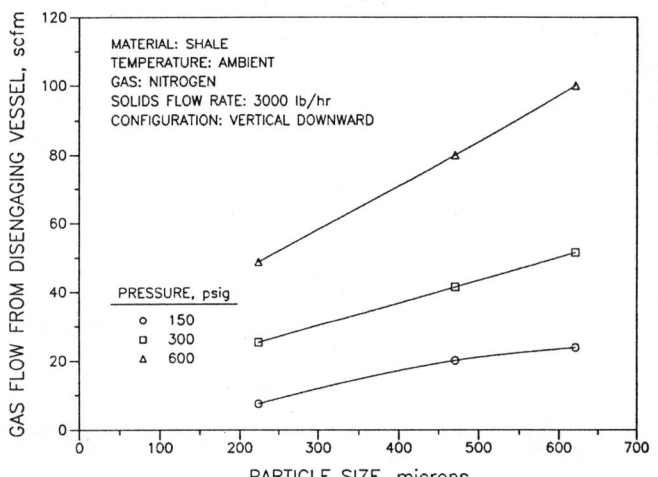

Figure 12. Effect of particle size on the gas flow rate from the disengaging vessel.

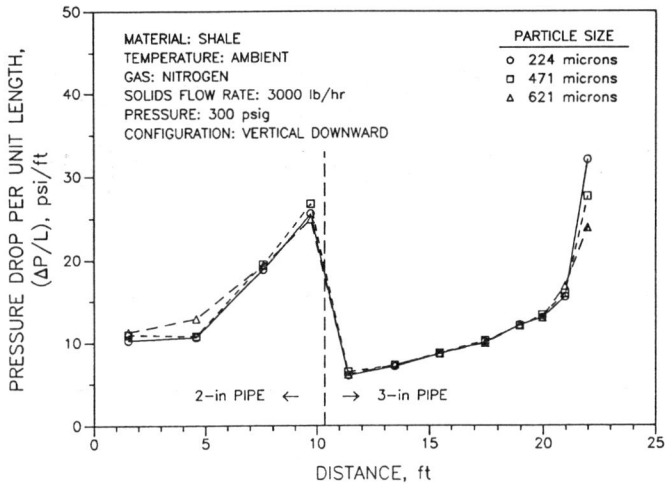

Figure 13. Effect of particle size on the pressure drop distribution.

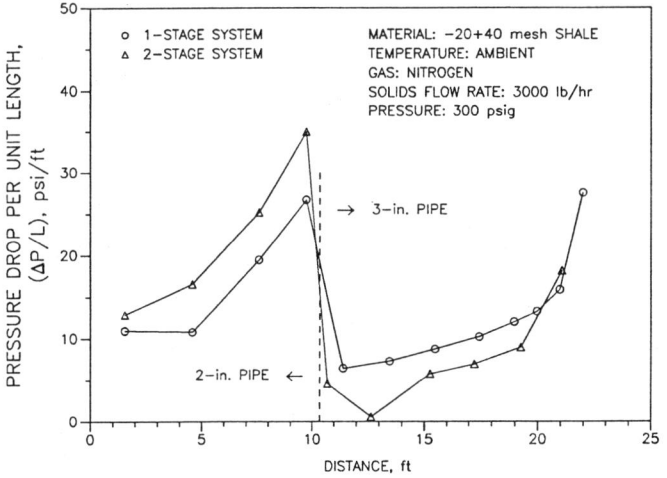

Figure 14. Comparison of pressure drop profiles for the two-stage and the single-stage systems.

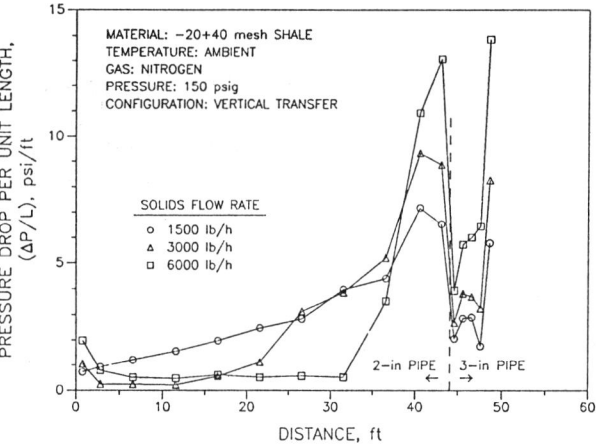

Figure 15. Pressure-drop profiles as a function of solids flow rate at a system pressure of 150 psig (vertical transfer system).

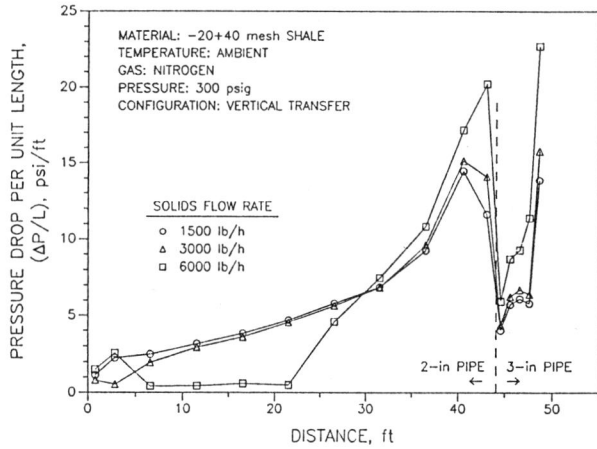

Figure 16. Pressure-drop profiles as a function of solids flow rate at a system pressure of 300 psig (vertical transfer system).

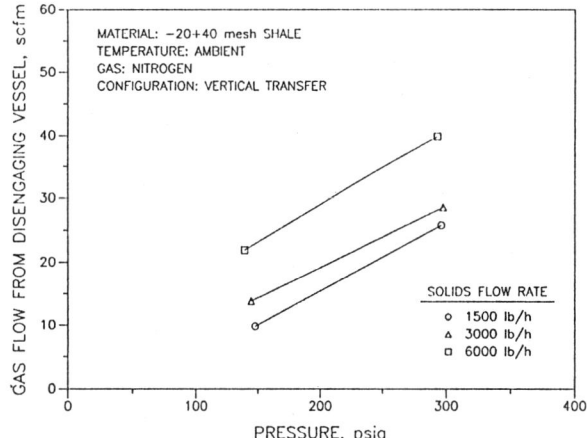

Figure 17. Gas flow from the disengaging vessel as a function of pressure (vertical transfer system).

AGGLOMERATION CLUSTER FORMATION OF FINE POWDERS IN GAS-SOLID TWO PHASE FLOW

H.O. Kono, T. Matsuda, C.C. Huang and D.C. Tian ■ Chemical Engineering Department, West Virginia University, Morgantown, WV 26506-6101

The agglomeration of dry, fine powders (Type A-C mixtures classified by Geldart (1973)) was experimentally observed in the gas-solid two phase flow. The size distribution of those agglomerates was measured in two dimensional transparent glass vessels by using a long distance focus microscope. The experimental results showed that the size distribution of powder agglomerates is determined by the gas-solid flow condition and the properties of the fine powders.

INTRODUCTION

Although several investigators have experimentally and theoretically studied gas-solid two phase flow, they have mostly used relatively coarse powders (Type A, B, or D Powders of Geldart's classification (1)). Therefore, they assumed that there was no powder agglomeration and that the powders were discretely dispersed in the gas phase. When fine or ultrafine powders or their mixtures are used, the powders generally form agglomerates. Recently the agglomerate formation of fine powders in gas-solid two phase flow regimes was studied and its results were reported; e.g. at an aerated or bubbling flow regime (2, 3, 4, and 5) also at an extremely high turbulent flow in turboclassifier (Kono et al. (1988)). The agglomeration defined above in the two phase flow is fundamentally different from the clusters defined by Yerushalmi et al. (6) which are substantially packets of coarse particles caused by particle slugging. In this study the agglomeration formation was experimentally confirmed in the fine powder-gas two phase flow system for the first time in open literature. The size and number of agglomerates were found to depend upon the gas velocity, the mass velocity ratio of powder/air, and the properties of powders involved. These findings are significant, since the formation of fine powder agglomerates strongly affect properties of the bulk powder in various powder handling processes.

EXPERIMENT

Fluid catalytic cracking (FCC) catalyst powers and aluminum hydroxide (Al(OH)3) powders were used as sample powders. The properties of these powders and their mixtures are shown in Table 1. Rheological properties of the bulk powders defined by Kono et al. (3, 4) are also provided in Table 1. Properties of the sample powders were controlled with mixing ratios of 0, 5, 10, and 15% by weight of Al(OH)3. The experimental apparatus is schematically shown in Figure 1. In order to precisely attain various flow conditions of the gas velocity (Uo), the mass flow ratio of powders and gases (Gs/Gg), and the properties of powders, a transparent rectangular glass vessel with one-pass gas-solid flow was used. The gas velocity range was 1 to 10 m/s. The range of the mass velocity ratio of powders and gases was 1 to 10. Therefore, the experimental ranges do cover turbulent, entrained, and/or fast flow conditions. The behavior of powders in two phase flow was observed by a high speed camera and TV, equipped with a long distance focus (50 to 60 mm) microscope. Its magnification was 50 to 200 times. The exposure time was 1/7700 sec. synchronized by a stroboscope. The size distribution of formed agglomerates was measured by the camera and TV explained above. The location of measurements was in the fourth unit from the bottom. The development of dynamically stable agglomerate size distributions was investigated through observation at various heights of the vessel.

RESULT AND DISCUSSION

Photograph 1 shows an image of powders in gas-powder two phase flow. The gas velocity was 2 m/s. Any particles moving slower than 2 m/s appeared as if they stopped. The range of image was approximately 2 x 2 mm. This example picture clearly shows that the powder flows in a very populated fashion and the flocculates are formed. As shown in Photograph 2 (a), the Al(OH)3 powders clearly coat the FCC particles. Photographs 2 (b) and (c) show the formation of FCC particle flocculates. When the particle travel faster then 2 m/s, particle motion can be recognized from the traces of trajectory. The velocity can be calculated from the length of the trajectory trace. The velocity of particles in Photographs 2 (d) and (e) is estimated as approximately 0.5 m/s.

The behavior of powders changes in accordance with the change of powder properties under a certain flow condition. Photograph 3 shows the behavior of powders (gas velocity 1 m/s, Gs/Gg = 2.0) for various mixing ratios of fine 0, 5, 10, and 15 % Al(OH)3 by weight. By increasing the mix ratio of Al(OH)3, the particle interaction was enhanced as reported by Kono et al. (3, 4, 1988, and 5).

Thus, the agglomerates grew in size when the fraction of Al(OH)3 fine powders increased. In other words, when the tensile strength of bulk powders increased, the agglomerates grew in size under the same hydrodynamic flow condition. Photograph 4 shows the images of agglomerates in two phase flow by increasing the gas velocity from 1 to 10 m/s.

The size distribution of agglomerated powders in two phase flow at the gas velocity 1 m/s is shown in Figure 2 for various mix ratios of Type C Powders, Al(OH)3. The cohesive property of the sample powders which was measured by the tensile strength (Table 1), increased by increasing the amount (0 to 15% by weight) of Type C Powder to FCC powder. As is seen in Figure 2 the median size of agglomerates increased from 5.5 micron to 12 micron under the specific flow condition (Uo = 1 m/s, Gs/Gg =2).

The corresponding distribution curve is shown in Figure 3 for a gas velocity Uo of 2 m/s. Although the median size of sample powders still increased with increased Type C Powder addition, the effect was not as remarkable as for the Uo = 1 m/s case (Figure 2).

The effect of the powder/gas mass velocity ratio (Gs/Gg) on the agglomerated powder size is shown in Figure 4 for the same powder at a constant gas velocity, Uo of 2 m/s. These results show that the concentration of agglomerates in the gas phase also seems to play an important role. From Figures 2, 3, and 4, the effect of the mass velocity ratio (Gs/Gg) and the properties of powders on the size of powder agglomerates was confirmed. Based on the experimental results, the size distribution of agglomerates seems to be determined by the dynamic balance between the cohesive force of the powder and the prevailing force of the specific flow condition. In order to verify the appropriate selection of the measurement height where the dynamic balance is reached, the size distribution of agglomerates was measured at various heights in the vessel. As shown in Figure 5 the size distribution of agglomerates at the center of the fourth unit from the bottom showed that the powder had reached its dynamic equilibrium condition.

CONCLUSION

The agglomerate formation of very fine powders in turbulent, entrained, and/or fast flow was observed, when FCC powders (30 to 100 microns) were mixed with Al(OH)3 (7 microns) at 5 to 15% (wt). The agglomeration depended upon fine powder rheological properties, mass velocity ratio of powder/gas (Gs/Gg), and gas velocity.

LITERATURE CITED

1. Geldart, D., Powder Technology, 7, 285 (1973).
2. Kono, H.O., et al., Powder Technology, 48, 51 (1986).
3. Kono, H.O., et al., Powder Technology, 52, 69 (1987a).
4. Kono, H.O., et al., Powder Technology, 53, 163 (1987b).
5. Kono, H.O., "Characterization of Fine Powders in the Emulsion Phase

of Fluidized Beds - Structural Model," Chapter I in *Transport in Fluidized Particle Systems*, edited by Doraiswamy, L.K., et al., Elsevier Science Publishers, B. V., Amsterdam, Netherland (1989).
6. Yerishamli, J., *Coal Processing*, vol. 3, pp.156, AIChE, New York (1977).

ACKNOWLEDGMENT

This work was partially supported by Central Research Institute of Electric Power Research Industry.
This work was also partially supported by U.S. National Science Foundation Grant No. CDR-87-15406.

Table 1. Physical Properties of the Powders.

Powders		particle diameter [micron]	particle density [Kg/m3]
FCC		70.6	1400
Al(OH)3		6.8	2400
FCC + Al(OH)3	5 wt%	67.2	1450
FCC + Al(OH)3	10 wt%	65.2	1500
FCC + Al(OH)3	15 wt%	62.3	1550

Powders		tensile strength of powder* [Pa]	bulk density [kg/m3]
FCC		80	795
Al(OH)3		930	-
FCC + Al(OH)3	5 wt%	97	763
FCC + Al(OH)3	10 wt%	110	745
FCC + Al(OH)3	15 wt%	150	742

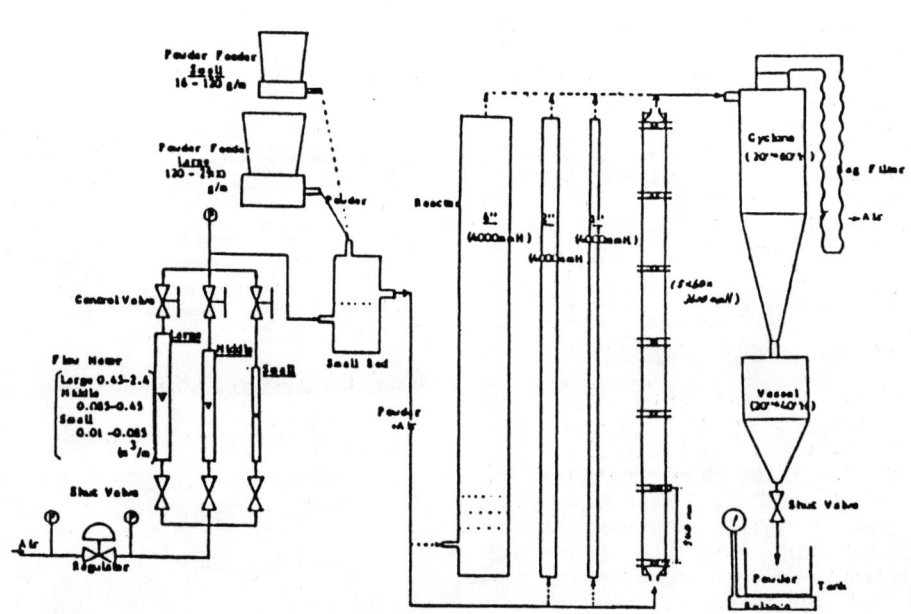

Figure 1. Schematic Diagram of Experimental Apparatus.

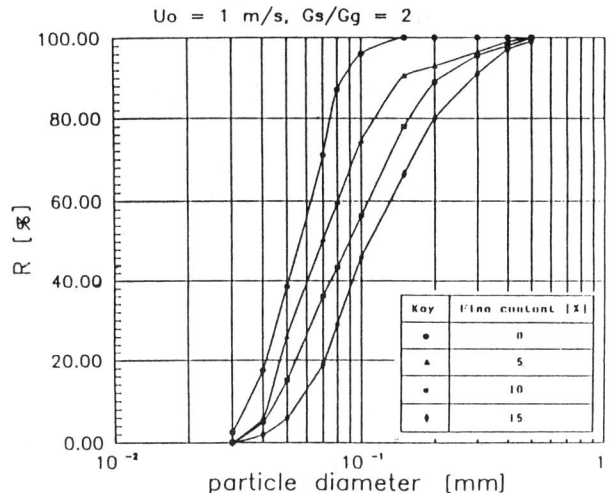

Figure 2. Size Distribution of Agglomerated Powders in Two Phase Flow Uo = 1 m/s, Gs/Gg = 2, Mixed Type C Powder Al(OH)3 0, 5, 10 and 15 wt.%.

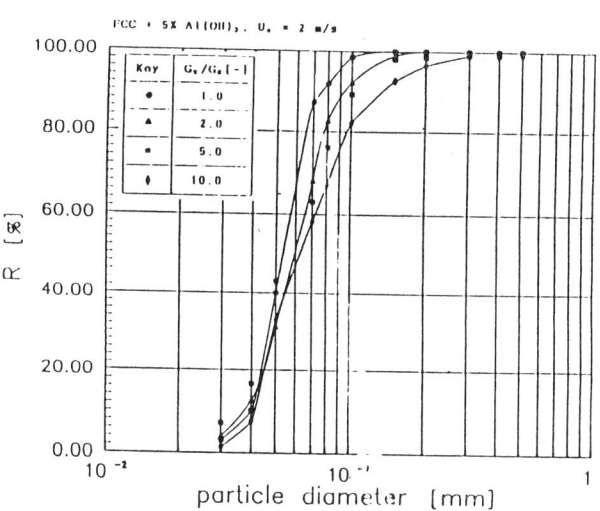

Figure 4. Size Distribution of Agglomerated Powders in Two Phase Flow Uo = 2 m/s, Gs/Gg = 1, 2, 5, and 10, Mixed Type C Powder Al(OH)3 5 wt.%.

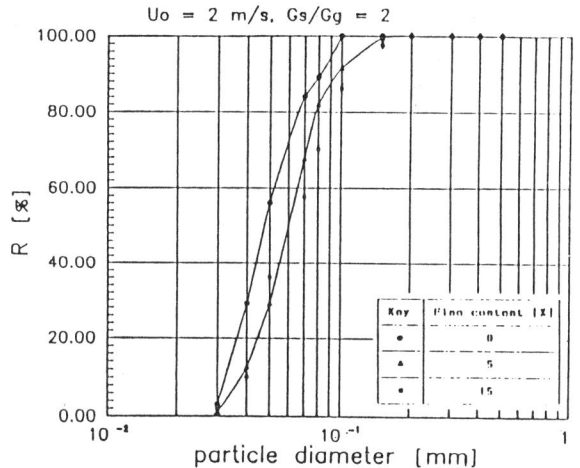

Figure 3. Size Distribution of Agglomerated Powders in Two Phase Flow Uo = 2 m/s, Gs/Gg = 2, Mixed Type C Powder Al(OH)3 0, 5, and 15 wt.%.

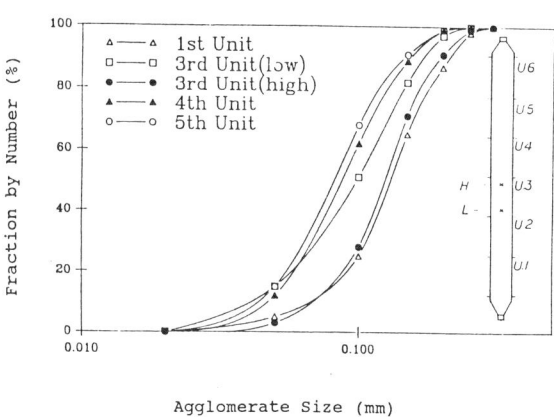

Figure 5. Size Distribution of Agglomerated at Various Heights U0 = 1 m/s, Gs/Gg = 3, Al(OH)3 5 wt.%.

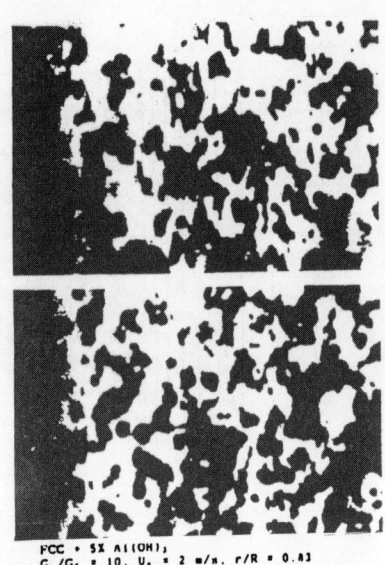

Photograph 1. Typical Image of Powders in Two Phase Flow Gas Velocity: 2 m/s Gs/Gg = 10. Size of Image: Approximately 2 mm x 2 mm. Shutter Speed: 1/7700 sec Stroboscope. FCC+5% Al(OH)3.

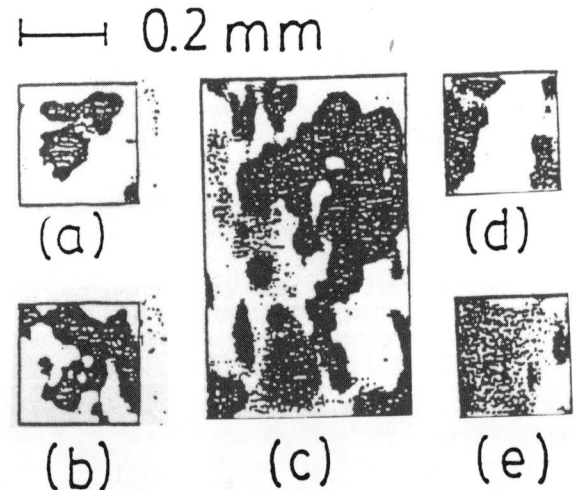

Photograph 2. Images of Long Distance Focus Microscopic Pictures (enlarged).
(a) FCC Powders with Fine Al(OH)3 Powders
(b) Agglomerated FCC and Al(OH)3 Powders
(c) Large Agglomerated FCC and Al(OH)3 Powders
(d) & (e) Fine Al(OH)3 Powders.

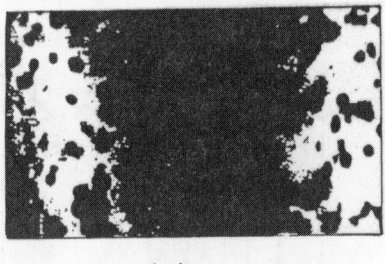

(a)

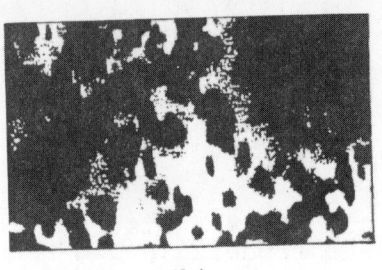

(b)

(c)

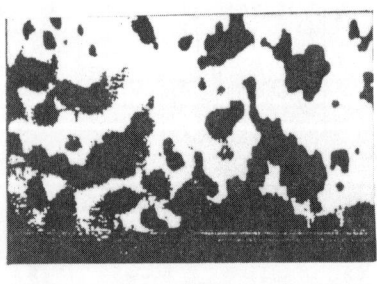

(d)

Photograph 3. Microscopic Pictures of Powders in Flow for the Different Concentration of Fine Al(OH)3 Powders Gas Velocity: 1 m/s, Gs/Gg = 2
(a) 0 wt.%, (b) 5 wt.%, (c) 10 wt.%, (d) 15 wt.%.

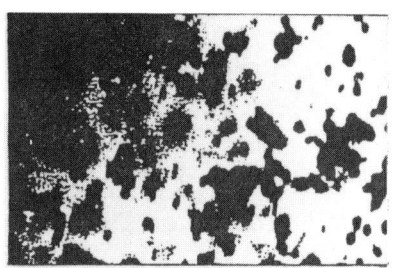
Uo = 1 m/s

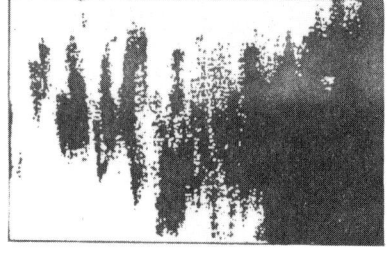
Uo = 5 m/s

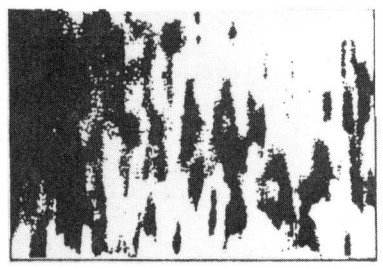
Uo = 2 m/s

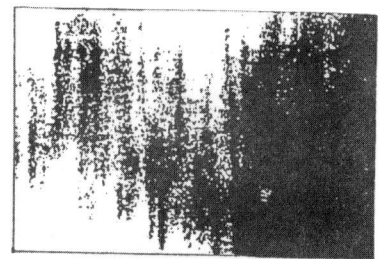
Uo = 10 m/s

Photograph 4. Images of Moving Particles. Gas Velocity: 1, 2, 5, and 10 m/s, Gs/Gg = 2 Type C Powder Al(OH)3 5 wt.%.

ANALYSIS OF GAS MOTION AT THE SURFACE OF A FLUIDIZED BED DUE TO BUBBLE ERUPTIONS

E.K. Levy and B. Kocatulum ■ Energy Research Center, Lehigh University, Bethlehem, PA 18015

A numerical analysis of the motion of the emulsion and gas phases driven by a single kidney-shaped bubble rising in a fluidized bed has been performed. The analysis is an extension of the Davidson model for a spherical bubble, modified to account for the nonspherical shape resulting from the presence of the bubble wake. Additional equations were derived to account for the upward motion of the bubble and the presence of the free surface. A combination of finite element and finite difference methods was implemented to solve the resulting set of equations, and solutions are presented for particle and gas velocities as functions of the vertical position of the bubble. Comparison of the results to previous results obtained for the eruption of a spherical bubble show that the gas velocity profiles are sensitive to the presence of the wake region. As a result, the wake region must be included when calculating gas flow patterns through an erupting bubble.

INTRODUCTION

When bubbles erupt from the surface of a gas-fluidized bed, they cause particles to be ejected into the freeboard. Some of these particles have a large enough initial momentum to be carried away from the bed with the aid of the upward gas flow. The rate of elutriation also depends strongly on the variation of the local gas velocities at the free surface arising due to the eruption of the bubbles. In previous investigations of the bubble eruption process, Do, et al. ([1]), Levy, et al. ([2]) and others have measured the particle motion in the freeboard resulting from the eruption of individual bubbles. In addition, Pemberton and Davidson ([3]), Y. Levy and F. Lockwood ([4]), Caram, et al. ([5]), Yule and Glicksman ([6]), Baskakov, et al. ([7]) and Levy, et al. ([8]) have all investigated different aspects of the transient gas flow field at the free surface resulting from the eruption of individual bubbles. In their study, Levy, Chen, Radcliff and Caram ([8]) developed a mathematical model for the motion of the emulsion and gas phases as a bubble moves upward through a fluidized bed and erupts at the surface. The results, which were presented for the case of a spherical bubble, show that the gas velocities across the free surface and through the rising bubble increase when the bubble approaches the free surface. When the bubble erupts, the flow pattern of gas and particles and the shape of the free surface change substantially.

The present investigation extends the analysis of Reference ([8]) by adding a wake region to the bubble, thus treating the bubble as a non-spherical symmetric cavity. The gas and solids flow fields associated with the eruption of a bubble having a wake fraction of 0.18 are described in the paper. Comparisons are made to the previous results for the spherical case, showing the sensitivities of the gas flow patterns to the presence of the wake region.

THEORETICAL APPROACH

The analysis begins with equations developed by Davidson ([9]), who derived equations for the emulsion and gas flow fields associated with the rise of isolated single bubbles in a bed of infinite extent. He assumed that the bubble contains no solid particles and has a circular shape (i.e. cylindrical in the two dimensional case and spherical in the three dimensional case). The motion of the solid particles around the rising bubble was taken as that of an inviscid incompressible fluid. Davidson further assumed that the gas flows through the emulsion phase as a viscous incompressible fluid, thus obeying Darcy's law for flow through a porous medium.

In the present paper, the bubble is three dimensional and non-spherical, with a wake which occupies its lower portion. The analysis begins after the bubble has been formed. The shape of the bubble is specified at the beginning of the calculations and its shape and size remain fixed during the course of the motion as the bubble rises with constant velocity towards the free surface.

Following Davidson, the continuity equation for the particulate phase is

$$\nabla \cdot v_p = 0 \qquad (1)$$

Introducing the velocity potential for the particle motion ϕ results in

$$\nabla^2 \phi = 0 \qquad (2)$$

The continuity equation for the interstitial gas flow is

$$\nabla \cdot u_g = 0 \qquad (3)$$

and the relative velocity between solid particles and the gas phase is given by Darcy's law:

$$u_g - v_p = -k\nabla p \qquad (4)$$

where k is a constant. Taking the divergence of Equation (4) and combining with (1) and (3) this becomes

$$\nabla^2 p = 0 \qquad (5)$$

The solution of these equations begins with the bubble located deep in the bed, far below the free surface. Initially, the free surface is assumed to be flat and at rest. As the bubble rises, it causes the free surface to deform, with changes in shape which can be calculated by applying dynamic and kinematic conditions. From surface wave theory (Reference 10) the equation of the free surface, referring to Figure 1, is

$$z_f - \eta - z_0 = 0 \qquad (6)$$

Since the surface moves with the particles

$$\frac{D}{Dt}(z_f - z_0 - \eta) = 0 \qquad (7)$$

which, for an axisymmetric problem, reduces to

$$v_z = \frac{\partial \eta}{\partial t} + v_r \frac{\partial \eta}{\partial r} \qquad (8)$$

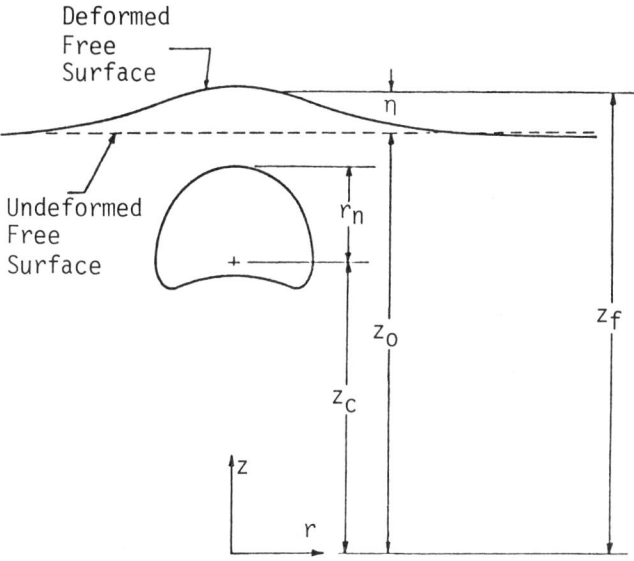

Figure 1. Coordinate System and Parameters Used in Governing Equations

Another boundary condition which accounts for the surface dynamics is obtained by applying the unsteady Bernoulli equation to the free surface

$$p_f = \rho_s(1-\varepsilon_{mf})\{(\frac{\partial \phi}{\partial t} - g(z_0+\eta) - \frac{1}{2}v^2) + C\} \qquad (9)$$

Since at the beginning of the process the free surface is flat and at rest:

$$t = 0 \qquad \eta = 0, \frac{\partial \phi}{\partial t} = 0, \quad v_p = 0$$

and Equation (9) gives

$$p_f = \rho_s(1-\varepsilon_{mf})[-gz_0 + C] \qquad (10)$$

Combining Equations (9) and (10), and rearranging, the dynamic boundary condition can be expressed as

$$\frac{\partial \phi}{\partial t} = g\eta + \frac{1}{2}v^2 \qquad (11)$$

Equations (2), (4), (5), (8), and (11) can be rendered non-dimensional by using the following reference parameters:

u_b : bubble velocity

r_n : bubble nose radius (Figure 1)

$\rho_s(1-\varepsilon_{mf})r_n g$: pressure

The resulting non-dimensional equations are:

$$\nabla^2 \Phi = 0 \quad (12)$$

$$\nabla^2 P = 0 \quad (13)$$

$$U_g = V_p - K\nabla P \quad (14)$$

$$\frac{\partial H}{\partial T} = V_z - V_r \frac{\partial H}{\partial R} \quad (15)$$

$$\frac{\partial \Phi}{\partial T} = \frac{H}{1.5828 Fr} + \frac{1}{2}(V_r^2 + V_z^2) \quad (16)$$

where K is a dimensionless interstitial gas velocity,

$$K = \frac{u_{g\infty}}{u_b}$$

and Fr is the Froude number defined by

$$Fr = \frac{u_b^2}{g\, d_{eq}}$$

The constant in Equation (16) comes from the relation between the nose radius and the equivalent diameter of the bubble selected for analysis:

$$d_{eq} = 1.5828\, r_n$$

Note that this coefficient is unique to the bubble shape shown in Figure 1. Bubbles having other shapes and wake fractions will result in different values for this parameter. Finally for an isolated bubble, the bubble rise velocity is

$$u_b = 0.711\, (g\, d_{eq})^{1/2} \quad (17)$$

The time dependent portion of the problem was solved by employing the finite difference technique. Using a forward-difference for the time derivative and a central difference for the space derivative results in the following implicit scheme for Equation (15):

$$\frac{H_i^{n+1} - H_i^n}{\Delta T} + V_{r_i}^n \frac{H_{i+1}^{n+1} - H_{i-1}^{n+1}}{2\Delta R} = V_{z_i}^n \quad (18)$$

where subscript i stands for the spatial points and superscript n refers to the levels of time separated by the interval ΔT.

Rearranging Equation (18) results in

$$b_i H_{i-1}^{n+1} + d_i H_i^{n+1} + a_i H_{i+1}^{n+1} = c_i \quad (19)$$

where

$$b_i = -\Delta T\, V_{r_i}^n$$

$$d_i = 2\, \Delta R$$

$$a_i = \Delta T\, V_{r_i}^n$$

$$c_i = 2\, \Delta R\, (H_i^n + \Delta T\, V_{z_i}^n)$$

Applying the symmetry condition for the free surface at the centerline

$$\frac{\partial H}{\partial R} = 0 \quad \text{at } i = 0$$

In addition, at the free surface and far from the bubble

$$\frac{\partial H}{\partial R} = 0 \quad \text{at } i = M$$

These two conditions are satisfied provided that

$$a_0 = 0 \quad \text{and} \quad b_M = 0$$

The system of equations given by Equation (19) was solved by applying the Thomas algorithm (Reference 11). A first order explicit scheme was employed to solve the velocity potential equation.

$$\Phi_i^{n+1} = \Phi_i^n + \Delta T \left(\frac{H_i^n}{1.58 Fr} + \frac{1}{2}\{(V_{r_i}^n)^2 + (V_{z_i}^n)^2\} \right) \quad (20)$$

At any time level n, the boundary conditions for the particle velocity potential are

free surface $\quad : \Phi = \Phi^n$

surface of bubble $\quad : \frac{\partial \Phi}{\partial n} = -\cos\theta$

bottom edge $\quad : \frac{\partial \Phi}{\partial Z} = 0$

centerline and right edge : $\frac{\partial \phi}{\partial R} = 0$

The pressure field was obtained by solving Equation (13) subject to the following boundary conditions:

free surface : $P = 0$

surface of bubble : $P = P_b$

centerline : $\frac{\partial P}{\partial R} = 0$

bottom and right edge : $P = Z_0 + H - Z$

where P_b, the pressure inside the bubble, is a function of time only. This is based on the assumption that at any instant of time, the spatial variations of pressure within the bubble are very small compared to the variations in the surrounding emulsion phase.

The interstitial gas velocity is computed by using the solutions for the particle velocity and the pressure via Equation (14). For a given value of K, the gas velocity obtained from Equation (14) must satisfy conservation of mass applied to the bubble. Hence, it is necessary to go through an iterative procedure to determine the bubble pressure P_b for which the resulting gas flow field yields a zero net gas flow rate through the bubble.

SOLUTION ALGORITHM

The time dependent nature of the problem coupled with the fact that the solution domain has a different shape at each time level, necessitates a step-by-step process. In this case, the results of any given time level determine the shape of the domain and the boundary condition at the free surface of the bed for the next time step.

At each time step, Equations (12) and (13) were solved by the finite element method using the commercial software package, Twodepep, developed by International Mathematical and Statistical Libraries, Inc. (IMSL). Ten-node cubic triangular elements were used in the calculations. Here, care was taken to make the elements as close to an equilateral shape as possible, a condition desired for accuracy (Reference 12). The typical mesh dimension ℓ at the bubble surface was of order $\ell/r_n \sim 1/2$ for a bubble deep in the bed. The smallest mesh size was used in the bulge region for a bubble at the last stages of eruption. In this case $\ell/r_n \sim 1/8$ (see Figure 2 for an example of the mesh configuration used).

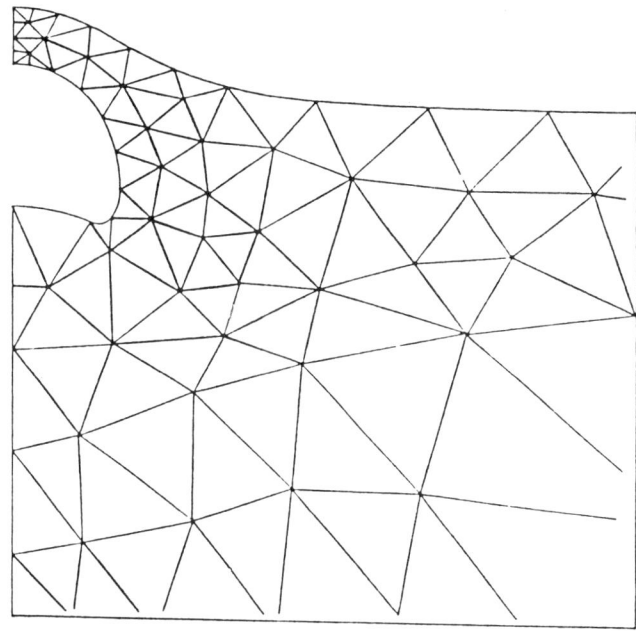

Figure 2. The Finite Element Mesh in the Vicinity of the Bubble for the Last Stages of the Computation

The gas flow field was computed from Equation (14) and the result was used to determine if conservation of mass through the bubble was satisfied. The pressure boundary condition on the surface of the bubble was modified, Equation (13) was solved again and this iterative procedure was continued until the resulting gas flow field yielded a zero net gas flow rate through the bubble.

The bubble was then moved upward a distance $\Delta z = u_b \times \Delta t$, and Equations (15) and (16) were employed to obtain the new shape of the free surface H^{n+1} and the new value of the particle velocity potential at the free surface ϕ^{n+1}. The size of the time increment ΔT was selected based on Chen's analysis (Reference 13) in which he tried different ΔT values and determined the time step size which gave stable solutions.

Once the new values of the deformation and the velocity potential were obtained at discrete points throughout the free surface, a least-squares curve fitting method was applied to express each of the two variables as a set of smoothly joined polynomials.

RESULTS

The results shown here are for a bubble with a wake fraction, $f_w = 0.18$ and with the specific bubble shape illustrated in Figure 1. A cylindrical domain with a diameter of 40 r_n and a depth of 40 r_n was selected based on Radcliff's analysis (Reference 15) in which he carried out calculations with different domain sizes and compared the results to those obtained from the Davidson theory. The initial location of the bubble was taken as $6r_n$ from the free surface of the bed. This starting point was selected based on Chen's (Reference 13) results which showed no significant effect of the bubble on the free surface until the bubble center was within 3 diameters of the free surface. The value of the Froude number was 0.50552. This corresponds to conditions near minimum fluidization and also to a deep freely bubbling bed, where the bubbles are sufficiently far apart from each other so that no interactions occur between them.

The results presented here are identified with the instantaneous vertical position of the effective bubble center, Z_c. This characterizes the time-variable for the transient process. The calculations were started at $Z_c = 34.0$ and terminated at $Z_c = 39.3$. During this interval, the bubble was moved toward the free surface in 14 separate time steps of decreasing step size Δt, with ΔZ_c varying from 1.0 to 0.1.

The region between the bubble and the free surface is critical in understanding the elutriation phenomenon. Figure 3 depicts the development of the dome at the free surface in relation to the location of the bubble. Once it starts, the deformation of the free surface progresses quite rapidly. Shown in a different way in Figure 4 is the deformation of the free surface as a function of radial distance from bubble centerline and the vertical location of the bubble. As the bubble rises, there is a considerable increase in the deformation of the free surface, especially when the bubble is within $2r_n$ of the free surface. The vertical deformation is a maximum at the centerline, with a steep decrease with radial distance until R=3, after which the vertical displacement becomes negative. These negative values are necessary in order to satisfy overall conservation of mass for the bed material. The particle velocities in the emulsion phase in the vicinity of the bubble are illustrated in Figure 5. The magnitude of the particle

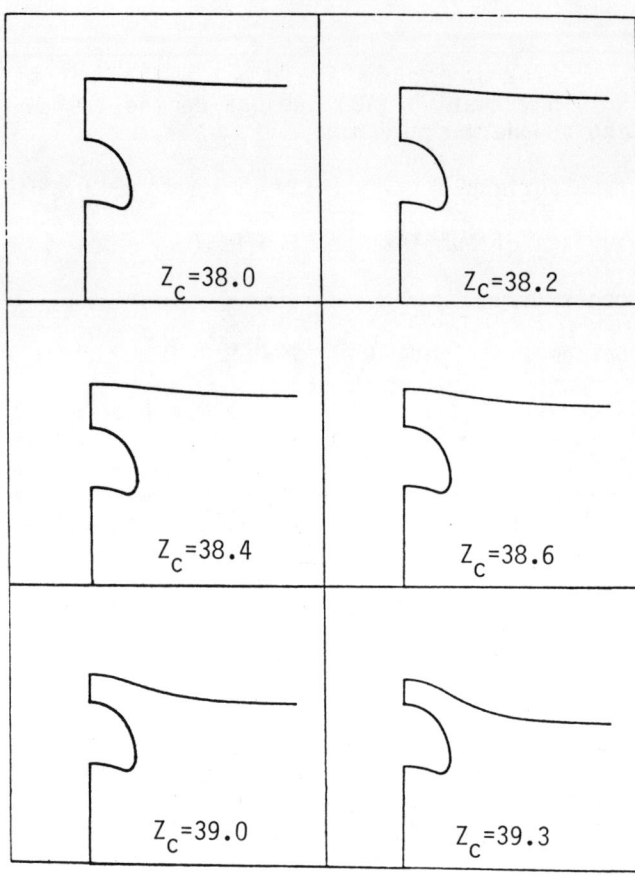

Figure 3. Shape of the Free Surface during Bubble Rise

velocity diminishes rapidly with distance away from the bubble. At the nose of the bubble, the particles are pushed upward, and the particulate phase moves into the region behind the bubble to fill the space being vacated by the bubble.

The gas flow in the vicinity of the bubble is presented in Figures 6 and 7. These graphs display the relative gas velocity with respect to the bubble for slow(K=10.0) and fast(K=0.01) bubbles. It should be noted that the vector scale in Figure 6 is different from that in Figure 7. The radial component of gas velocity is significant only in the immediate vicinity of the bubble, and in general, the vertical component dominates the gas flow field. In the case of a fast bubble (Figure 6), the gas flow is downward with respect to the bubble throughout the flow field except in the region close to the bubble where it exhibits a recirculating pattern. For slow bubbles

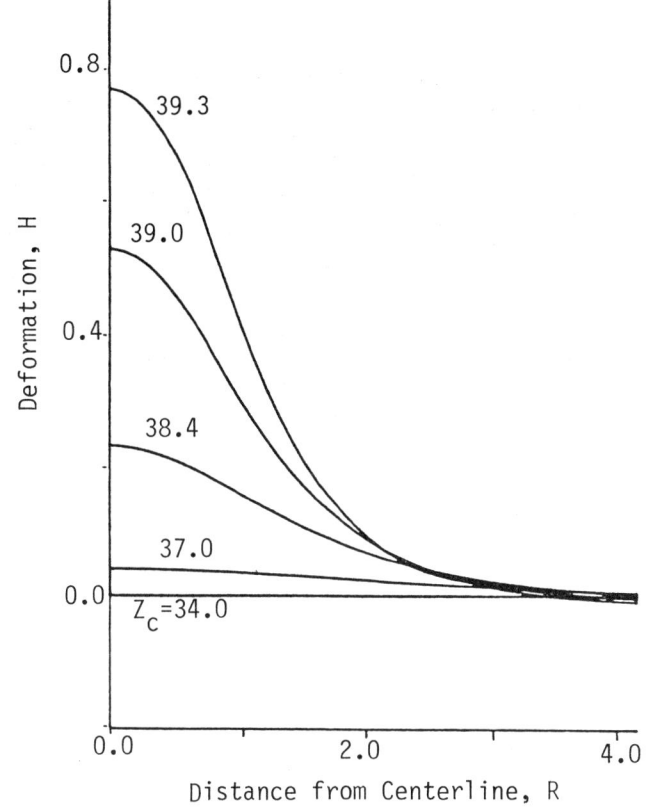

Figure 4. Deformation of Free Surface Highlighting Area in the Vicinity of Bubble Centerline

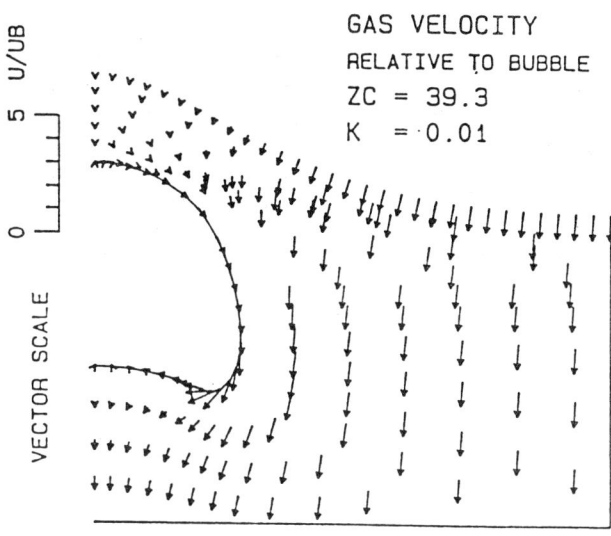

Figure 6. Vector Field of Interstitial Gas Velocity Relative to Bubble Highlighting Area near Bubble for $Z_C = 39.3$ and $K = 0.01$

(Figure 7), the gas flows upwards, entering the bubble from below and flowing out around the top. The large gradient in gas velocity within the bulge layer at the nose of the bubble shown in Figure 7 is due to the r^2 dependence of surface area which occurs in the case of a spherical geometry.

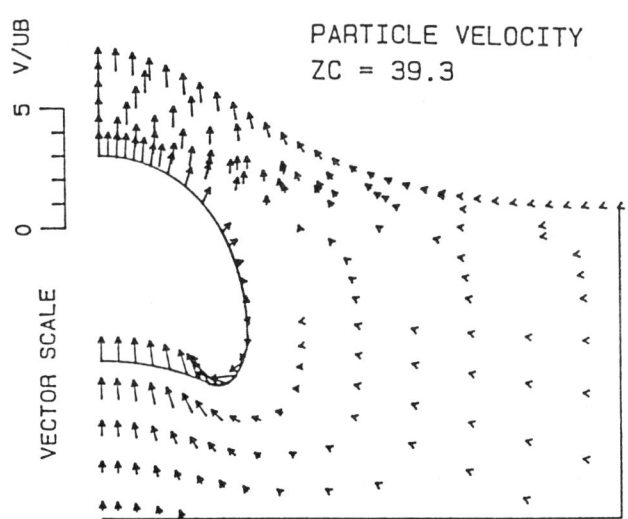

Figure 5. Vector Field of Absolute Particle Velocity Highlighting Area near Bubble for $Z_C = 39.3$

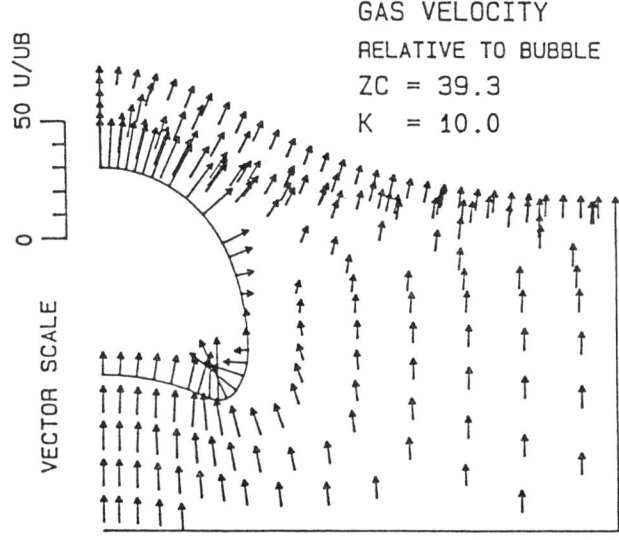

Figure 7. Vector Field of Interstitial Gas Velocity Relative to Bubble Highlighting Area near Bubble for $Z_C = 39.3$ and $K = 10.0$

In order to illustrate the characteristics of the gas motion at the free surface more clearly, the spatial and time-wise variations of the vertical component of the absolute interstitial gas velocity at the free surface is plotted in Figures 8 and 9 for K=0.01 and K=10.0. In the case of a fast bubble, the velocity profile has a characteristic shape with the maximum velocity at the centerline and a smooth decrease in velocity with radial distance (Figure 8). In the slow bubble regime, the velocity profile is more irregular, with the peak shifted off-center (Figure 9). As is seen from Equation (14), the gas velocity depends on the particle velocity field and on a pressure gradient term. In the case of a fast bubble with K=0.01, the pressure gradient term is negligible compared to the particle velocity, resulting in a variation of the vertical interstitial gas velocity which is almost identical to the variation of the vertical particle velocity at the free surface. In contrast, in the case of a slow bubble with K=10.0, the vertical component of the interstitial gas velocity at the free surface is dominated by the pressure gradient term, thus leading to the irregular shapes of the velocity profiles seen in Figure 9.

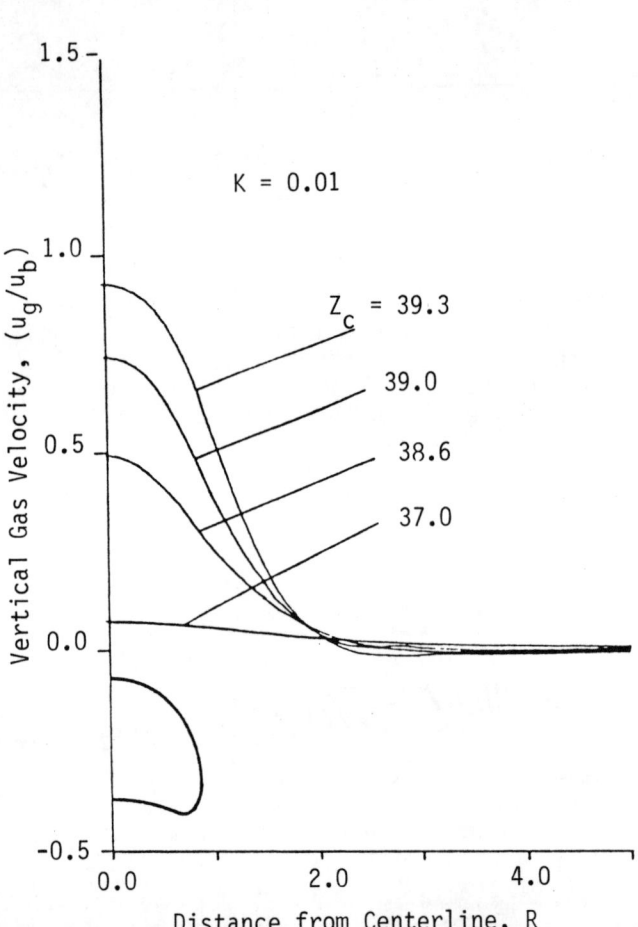

Figure 8. Vertical Component of Absolute Interstitial Gas Velocity at Free Surface for K = 0.01

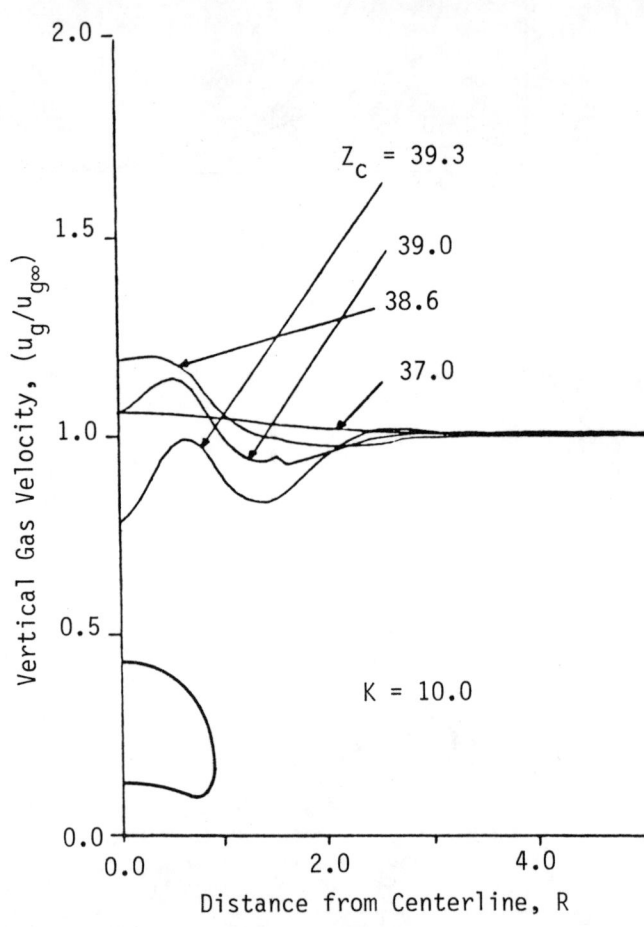

Figure 9. Vertical Component of Absolute Interstitial Gas Velocity at Free Surface for K = 10.0

DISCUSSION AND CONCLUSIONS

A comparison of the results obtained in this study with those published by Levy, et al. ($\underline{8}$) for a spherical bubble show some similarities but also some very important differences. The deformation of the free surface of the bed and the particle flow field in the vicinity of the bubble appear to be qualitatively similar for the case of the spherical and nonspherical bubbles. This is seen from Figure 10, which shows the change in the thickness of the bulge layer at the nose of the bubble as a function of time. Also included in the graph are experimental values taken from high speed video sequences of erupting three dimensional bubbles (Reference 2).

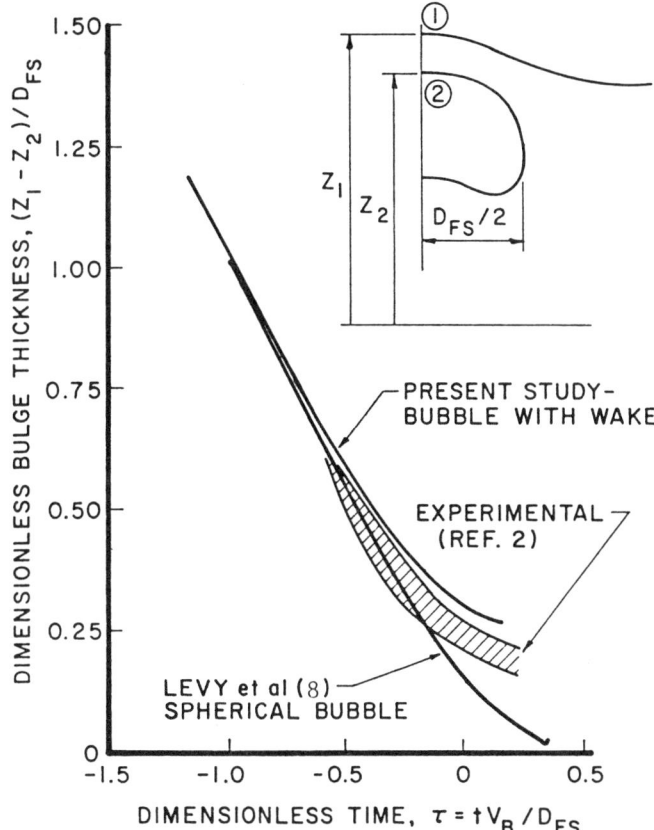

Figure 10. Decrease in bulge thickness with time during bubble eruption. Time $\tau = 0$ corresponds to the instant when the bubble nose (point 2) reaches the undisturbed level of the free surface of the bed.

It is the gas velocities which show the major differences between the two sets of calculations. In the case of the nonspherical bubble with K=0.01 (Figure 8), the vertical component of gas velocity at the free surface is positive for all values of the radial coordinate R and it increases towards a value $u_g/u_b=1.0$ at the bubble centerline. The results for the spherical bubble (Reference 8) also indicate a local maximum at the bubble centerline with the velocity approaching an asymptotic value of $u_g=u_b$. But in addition, the gas flows downward over a range of values of R from 0.8 to 3.0.

In the case of a slow bubble with K=10.0, the results for the spherical bubble (Reference 8) have the local maximum in gas velocity at the bubble centerline, with values approaching $u_g/u_b=4.0$. The results shown in Figure 9 for the nonspherical bubble show the maximum in velocity near R=0.7, with a local peak velocity in the range of $u_g/u_b=1.2$.

The shapes of the velocity profiles shown in Figure 9 are partially consistent with experimental findings of Y. Levy and Lockwood (4) and Caram, et al. (5) who performed experiments with incipiently fluidized beds with single bubble injection. While the gas velocities and bubble sizes used in those experiments correspond to fast bubbles with values of K in the 0.2 to 0.4 range, both groups of investigators saw the evidence of the formation of vortex rings due to the eruption of individual bubbles at the free surface. In both cases, this vortex ring had a direction of rotation as illustrated in Figure 11 which is a type of motion which could arise due to velocity profiles having the shape of those shown in Figure 9.

Figure 11. Observed Direction of Rotation of Vortex Ring Formed during Eruption of Individual Bubble at Free Surface of Bed

Figure 12 shows the total gas flow rate through the bubble Q_B as a function of the distance of the bottom of the bubble cavity from the undisturbed free surface of the bed. The results from the present analysis, for K=10, show a maximum flow rate occurring with $B/d_{eq} \approx 1.2$. Comparable results for a spherical bubble, obtained from Reference 8, show the peak flow rate occurring at B/d_{eq}

≈ 0.7. The results from Reference 6 indicate a peak value of $\Omega \approx 4.5$ occurring at $B/d_{eq} \approx 0.8$.

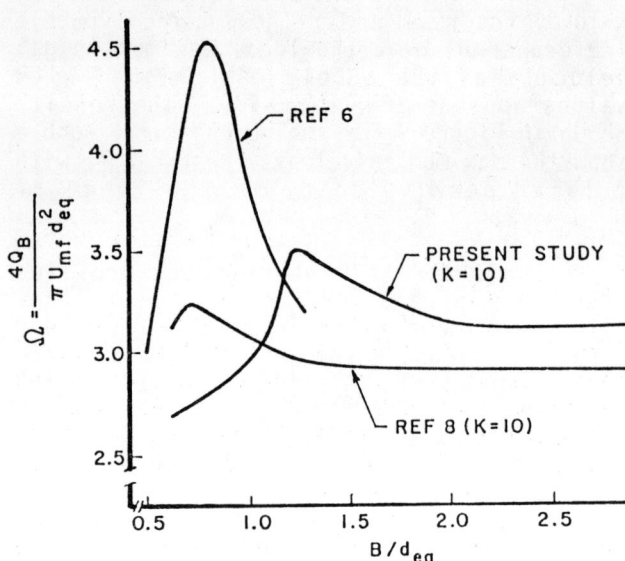

Figure 12. Variation of Gas Flow Rate through bubble with Vertical Distance from Bottom of Bubble Cavity to Undisturbed Free Surface of Bed

The results contained in this paper and in Reference 8 show that the presence of the wake region cannot be neglected in calculating the gas flow rates through an erupting bubble. Bubble cavities with flattened bottoms have higher peak flow rates than bubbles with spherical bottoms. In addition the flow rate is extremely sensitive to bubble aspect ratio, being greater for vertically elongated bubbles than for spherical or flat bubbles (References 6 to 8).

Finally, as shown in Reference 8, for a bubble of a given shape, the gas flow rate through the bubble is controlled primarily by the flow resistance provided by the bed material in the region surrounding the sides and bottom part of the bubble. These results suggest that during the last stages of eruption, the bulge layer has only a minor effect on the gas flow rate through the bubble. However, interactions between the gas and the particles in the bulge layer can be expected to have an important effect on the detailed transverse distributions of the transient gas velocity profiles at the top of the erupting bubble.

NOTATION

B	= Vertical distance from bottom of bubble cavity to undisturbed free surface of bed
C	= Constant
d_{eq}	= Equivalent diameter of bubble; $d_{eq}=(6 V_b/\pi)^{1/3}$
Fr	= Froude number; $Fr=u_b^2/gd_{eq}$
g	= Gravitational acceleration
H	= Dimensionless free surface displacement; $H=\eta/r_n$
k	= Permeability constant
K	= Dimensionless interstitial gas velocity far from the bubble; $K=u_{g\infty}/u_b$
n	= Normal to surface
$n, n+1$	= Consecutive levels of time
p	= pressure
p_f	= pressure at free surface of bed
P	= Dimensionless pressure; $P=p/[\rho_s(1-\epsilon_{mf})r_n g]$
P_b	= Dimensionless pressure inside the bubble
Q_B	= Gas flow rate through bubble
r	= Radial coordinate
r_n	= Nose radius of bubble
R	= Dimensionless radial coordinate; $R=r/r_n$
t	= time
T	= Dimensionless time; $T=tu_b/r_n$
u_b	= Bubble rise velocity
u_g	= Interstitial gas velocity
$u_{g\infty}$	= Interstitial gas velocity far from the bubble
u_{mf}	= Minimum fluidization velocity
U_g	= Dimensionless interstitial gas velocity; $U_g=u_g/u_b$
v_p	= Particle velocity
v_r	= Radial component of particle velocity
v_z	= Vertical component of particle velocity
V_b	= Dimensionless volume of the bubble
V_p	= Dimensionless particle velocity; $V_p=v_p/u_b$
z	= Vertical coordinate
z_c	= Vertical position of the effective bubble center
z_f	= Height of free surface
z_0	= Level of surface of undisturbed bed
Z	= Dimensionless vertical coordinate; $Z=z/r_n$
Z_c	= Dimensionless vertical position of the effective bubble center; $Z_c=z_c/r_n$
Δt	= Time interval
ϵ_{mf}	= Voidage of the bed at minimum fluidization
η	= Deformation of the free surface
Ω	= Dimensionless gas flow rate; $\Omega=4Q_B/\pi U_{mf}d_{eq}^2$

ϕ = Velocity potential of emulsion phase
Φ = Dimensionless velocity potential; $\Phi = \phi/u_b r_n$
ρ_s = Density of the solids

LITERATURE CITED

1. Do, H. T., J. R. Grace and R. Clift, Powder Technology, 6, 195-200 (1972).

2. Levy, E. K., J. C. Dille and H. S. Caram, Powder Technology, 32, 173-178 (1982).

3. Pemberton, S. T. and J. F. Davidson, Chem. Eng. Sci., 39, 829-840 (1984).

4. Levy, Y. and F. C. Lockwood, AIChE Journal, 29, 889-895 (1983).

5. Caram, H. S., Z. Efes and E. K. Levy, "Gas and Particle Motion Induced by a Bubble Eruption at the Surface of a Gas Fluidized Bed", AIChE Symp. Series, 234(80), 106 (1984).

6. Yule, T. W. and L. R. Glicksman, "Particle Ejection from the Surface of a Fluidized Bed by an Erupting Bubble", Fluidization VI, Grace, J., L. Shemilt and M. Bergougnou (Ed.), Engineering Foundation (1989).

7. Baskakov, A. P., et al, Chemical Engineering Science, 39(3), 407 (1984).

8. Levy, E. K., H. K. Chen, R. Radcliff and H. S. Caram, Powder Technology, 54, 45-57 (1988).

9. Davidson, J. F., and D. Harrison, Fluidised Particles, Cambridge University Press, London (1963).

10. Milne-Thomson, L. M., Theoretical Hydrodynamics, Macmillan, 5th ed. (1968).

11. Anderson, D. A., J. C. Tannehill and R. H. Pletcher, Computational Fluid Mechanics and Heat Transfer, Hemisphere Publishing Co., New York (1984).

12. Huebner, K., Finite Element Method for Engineers, John Wiley and Sons, New York (1975).

13. Chen, H. K. "Gas Flow Through Bubbles in a Fluidized Bed", Master's Thesis, Lehigh University (1986).

14. Kocatulum, B. "A Numerical Study of the Fluid Mechanics within a Bubbling Fluidized Bed", Master's Thesis, Lehigh University (1989).

15. Radcliff, R. "A Mathematical Model of a Bubble Approaching the Free Surface of a Fluidized Bed", Master's Thesis Lehigh University (1984).

VIBRO-FLUIDIZATION OF GROUP-C PARTICLES AND ITS INDUSTRIAL APPLICATIONS

S. Mori, A. Yamamoto, S. Iwata, T. Haruta and I. Yamada ■ Nagoya Institute of Technology, Gokiso, Showa, Nagoya, 466, Japan

E. Mizutani ■ Chuo-Kakoki Co. Ltd., Nakanowari, Toyoake, Aichi, 470-11, Japan

To fluidize Geldart's group-C particles, a new type of vibro-fluidized bed has been developed.

A wide range of fine particles down to the submicron level can be fluidized fairly well at relatively lower gas velocities by this vibro-fluidized bed.

It is found that the fluidizability of these particles is strongly dependent upon the adhesive properties of the particles and the group-C particles can be classified into three sub-groups according to their fluidizability.

Some successfully developed industrial applications of the drying and humidity controls for the particles are presented. Other underdeveloped operations are also briefly discussed.

INTRODUCTION

Recently the handling of very fine particles including submicron particles is extremely important for the processing new materials, such as ; ceramics, plastics, metals, their composite materials, foods and drugs. However, these fine particles, classified into goup-C by Geldart (1973), have been believed to be unsuitable materials for fluidization since they channel easily in the bed and prevent good fluidization.

Morooka et al (1988) reported that even submicron particles were able to be smoothly fluidized since these particles agglomerated into larger particles and these secondary particles were easily fluidized at higher gas velocity. Although fine particles can be easily fluidized in the small bed, 3.5 cm I.D. bed was used by Morooka etal, their apparent minimum gas velocity for fluidization is increased with increasing bed diameter and a large amount of particles is entrained at a high gas velocity.

In this work, a new type of vibro-fuidized bed cold model of 20.5 cm I.D. is used to explain the fluidization behavior of different kinds of fine particles.

Some examples of successfully developed new industrial applications of this vibro-fluidized bed are also demonstrated.

EXPERIMENTAL METHOD FOR THE FLUIDIZATIN OF FINE PARTICLES

A cold model of the vibro-fluidized bed is schematically shown in Fig. 1.

The transparent plastic fluidized bed is 0.205 mID and 0.5m height.

The distributor is a porous plastic plate. Entrained particles from the bed are collected by a paper air filter and then weighed. The fluidized bed and its windbox are fixed on the upper surface of the rectangular vibro-stand which is supported on the base by four springs. A pair of vibro-motors are crosswisely fitted on the confronted side walls of the vibro-stand. The amplitude of the vibration can be changed by changing the angle of the impetus plates of the vibro-motors. The frequency is controlled by the inverter. The vibration angle of the bed is selected by changing the fitting angle of the vibro-motors.

As shown in Table 1, eight different kinds of particles and seven different size of alumina particles are used in this work. The size distribution of alumina particles are very narrow as shown in Fig. 2. In Fig.

3, particles are shown on the map proposed by Geldart (1973). Accurate densities of particles are not obtained and values of $\rho_P = 2\rho_b$ are plotted on this figure. It can be found from this figure that the particles used here are widely spread from 0.15 to 180μm. #240 to 700 alumina and PS particles are classified as group-A or B particles and other particles are group-C particles. In comparson with the original Geldart map (which covers the size of the particles larger than 10μm), the group-C particles used here (except alumina #1500) might be classified into extra fine C particles which are strongly adhesive and unsuitable for fluidization.

Commercial nitrogen gas is used as the fluidizing gas and the particles are well dried before each experimental run.

Bed collapse data are obtained by shutting off the flow of fluidizing gas only under vibrating conditions since irregular resonant vibration of the apparatus generated after cutting down the power of the vibro-motors disturbs the smooth bed collapse.

EXPERIMENTAL RESULTS

At first, the effect of the vibration agle is exarmined. When horizontal vibration is generated by keeping the angle of the vibro-motors at 0 rad, the unfluidized body of the bed is rotated slow and most of fluidizing gas blows up through the air gap that appers between the body and the surrounding wall.

Under vertical vibration at a $\pi/2$ rad motor angle, the air gap is generated between the body of the bed and the distributor plate and gas blows up through channels from the air gap. It is found that $\pi/4$ rad for the vibration angle is the optimum angle of the vibration to prevent the generation of channeling and to smoothly fluidize the fine particles. Therefore, all test data presented in this paper are obtained at this optimum vibration angle.

The bed expansion and fluidization states for different size alumina particles at 50 Hz frequency and 50mm amplitude are compared with that obtained under the non-vibrated condition in Fig. 4. In this figure, dashed lines show the partially fluidized states or channeling in the bed and solid lines show complete fluidization of the whole bed. It is shown in this figure that even increasing gas velocity up to 4.5 cm/s, particles smaller than 24μm cannot be completely fluidized under non-vibrated conditions. However, the fluidizability of the fine particles is improved by vibration of the bed and particles larger than 1μm can be completely fluidized at 1.8cm/s. 0.4μm particles can be completely fluidized at 4.5 cm/s. The bed expansion for 10μm particles is more than a factor 2, however for decreasing particle sizes below 10μm, bed expansions are decreased considerably.

Fig.5 shows the elutriation rate of various sized alumina particles at 4.5 cm/s. It is found from this figure that the elutriation rate is increased with decreasing particle size down to 5μm and then the rate is sharply decreases in further decreasing particles size.

The observed bed expansions for different kinds of particles obtained under a constant vibrating frequency of 40Hz and an amplitude of 0.5mm are shown in Fig. 6. Activated carbon and alumina #1500(10μm) particles show the largest bed expansion more than a factor of 2 and these particles are fluidized under almost bubble-free homogeneous conditions. The bed expansion of TiO_2(A100), SiO_2(FPS-1, CP100) and Aumina(AES-11) is 1.2-1.7 and they can be well fluidized like group-A particles.

Although the bed expansion of $MgCO_3$ (AM-50) particles is similar to CP100, it cannot be completely fluidized under this operating condition and a much higher gas velocity is required for complete fluidization since agglomerated particles larger than 1mm are easily formed in the bed.

$CaCO_3$(CC) particlesis are more adhesive and too difficult to fluidize and stronger vibration (60 Hz frequency and 1.25 mm amplitude) and a higher gas velocity (more than 10cm/s) are required.

Fig.7 shows the effect of the frequency of the vibration on the bed expansion of 1μm alumina particles as a typical example. The region to the left side of the cross-hatched line shows the region of incomplete fluidization. Solid lines are equi values of bed expansion. In this figure, the optimum frequency can be found for good fluidization at lower gas velocity. If 40 Hz is selected, the bed is completely fluidized at 1 cm/s gas velocity and the bed expansion becomes more than 1.5 at 2.5 cm/s. This effect of the vibration on the fluidizability of the fine particles is also dependent on the properties of the particles. Here, one of most impotant conclusions shoud be emphasised that the values of the minimun fluidization velocity and the terminal velocity of these fine particles cannot describe any of the fluidization and elutriation phenomena.

The experimental results for the collapse of the vibro-fluidized bed are shown in Figs. 8 and 9. Since these data are obtained under vibrated conditions and the final collapsed bed is compressed significantly, the final bed height becomes much smaller than the initial bed height. The final bed height here is selected as the static bed height (Lc). In these figures, the dashed line shows the experimental results for the alumina #360 (48μm) (which is the typical group-A particle) obtained under the non-vibrated fluidization.

It is found from these figures that most of the fine particles except $CaCO_3$ (CC) can be fluidized better than non-vibrated 48μm alumina particles. The expansion of emulsion phase for activated carbon (AC), alumina #1500 (10μm) and 3000 (5μm) are extraordinary large values.

It is concluded from the observed results shown in Figs. 6, 8 and 9 that the fine particles of group-C can be classified into three sub-groups with respect to their vibro-fluidizability: the first group is particles (including activated carbon, 5 and 10 μm alumina particles) which are very easily fluidized and give large bed expansion and elutriation. The second group is very adhesive partices like as $MgCO_3$ and $CaCO_3$ which are still difficult to fluidize even using the vibro-fluidized bed. The third group is common fine particles which are fluidized under bubbling conditions similar to group-A particles in the vibro-fluidized bed and their elutriation can be easily controlled.

INDUSTRIAL APPLICATIONS OF VIBRO-FLUIDIZED BED

The fluidized bed has been widely applied to various proceses, however in earlier types of fluidized beds is very difficult to treat group-C particles and subsequently applications of fluidization with these fine particles are not sufficiently developed. Although the vibro-fluidized bed proposed in this paper is not suitable for a large scale process (since the bed body reads to be vibrated), this bed is useful to handle relatively small amounts of expensive fine particles.

The new vibro-fluidized bed has various advantages: such as 1) most of the fine and adhesive particles are completely fluidized under a low gas velocity, 2) the elutriation of the particles from the bed is easily controlled, 3) the residence time of gas and particles are widely changed, and the common advantages of the ordinary fluidized bed are also available: for example, a closed and compact system, good mixing of the particles, the homogeneous and good controllability of the bed temperature.

Based on these advantages, new industrial applications are being developed using test units. One of the test units for the drying and humidity control of particles is shown in Fig. 10. Either a 0.25 m ID plastic fluidized bed can be selected for cold tests or a 0.3 m ID stainless steel for hot tests. Some examples of the results of industrial the applications are given here.

In Fig. 11, data for the drying organic fine particles less than 30 μm is shown as a typical example of drying fine particles.

The gas velocity is kept at 4 cm/s during the initial wetted period of the particles and then the velocity is decreased to 2 cm/s to protect the loss of the dried particles by elutriation.

Fig. 12 is an example of the removal of an organic solvent from particles. The organic particles discharged from the centifugal separator contain about 40 % of aqueous methanol solvent. The data for methanol removal using the vibro-fluidized bed are compared with results obtained from an ordinary fluidized bed drier. The wet particles can be well fluidized by the humidic air in the vibro-fluidized bed, therefore the methanol can be removed prior to the water at a lower bed temperature. The methanol content in the particles can be decreased to 0.003 % compared to 0.34 % in the conventioned fluidized bed drier.

Recently, the humidity control of fine particles has been required to prevent scattering or to maintain material quality. Fig. 13 is an example of the data on humidity control as the water content in HPC (hydroxy propyl celulose) is increased from 2 to 6 %. HPC particles are difficult to efficiently humidify because a gel film is formed at the wetted surface of the particles. It is found in this figure that the humidity of the exit gas gradually approaches the equilibrium humidity of the particles. The effect of vibration is significant as it prevents channeling of the bed except in the initial dry period of the operation.

The homogeneous mixing between different kinds of fine particles is also not easy but it is a very important operation for material processing. Usually wet methods are used in these processes and after treatments like drying are required. To develop a new dry mixing process, 1 kg of 0.95 μm zirconia particles (whose density is 6050 kg/m^3) is to be mixed with 9 kg of 0.4 μm alumina particles (whose density is 3930 kg/m^3) in a stirred vibro-fluidized bed test unit. It is found from the scanning electro-microscope photographs of the sintered sample that the zirconia particles can be well dispersed between the alumina particles after three hours of operation.

For a true evaluation of the dispersion of the particles, adequate material testing is required.

To separate unburned carbon particles from fly ash discharged from a pulverized coal fired power plant, a new classification process is under development using a batch type small unit where the fine fly ash is elutriated out and the carbon particle is cocentrated in the retaind bed. In another application, a new high temperature reactor to produce silicon carbonate and nitride from carbonized rice hull is being developed as a bench scale unit. These data will be presented in near future. Recently, new applications related to the surface treatment of fine metal particles have been successfully developed. This data will also be presented later.

CONCLUSION

A new type of the vibro-fluidized bed has been developed to handle group-C fine particles under fluidized conditions. The following results have been obtained:
1) A wide range of fine particles (down to the submicron particles) can be fluidized fairly well by this vibro-fluidized bed at relatively low gas velocities. The fluidizability of these fine particles is strongly depedent on their adhesive properties.
2) Group-C particles can be classified into three sub-groups according to their fluidizability.
3) The elutriation of the fine particles can be controlled very easily.
4) An optimum frequency can be selected to obtain good fluidization at low gas velocity.
5) The vibro-fluidized bed proposed here shows great for application to various industrial processes.

NOTATION

d_p = particle size [m]
$\bar{d}_p$ = average particle size [m]
f = vibration frequency [Hz]
g = acceleration of gravity [m/s^2]
K^* = elutriation rate [kg/m^3]
L_c, L_f = static and fluidized bed height, respectively [m]
l = amplitude of vibration [m]
u_0 = superficial gas velocity [m/s]
ρ_b, ρ_p = bulk and particle density, respectively [kg/m^3]
ρ_f = gas density [kg/m^3]
ω = angular frequency [rad/s]

LITERATURE CITED

Geldart, D.: Powder Technol., 7, 285 (1973)
Morooka, S., K.Kusakabe, A. Kobata, Y.Kato:
    J. Chem. Emg. Japan, 21, 41 (1988)

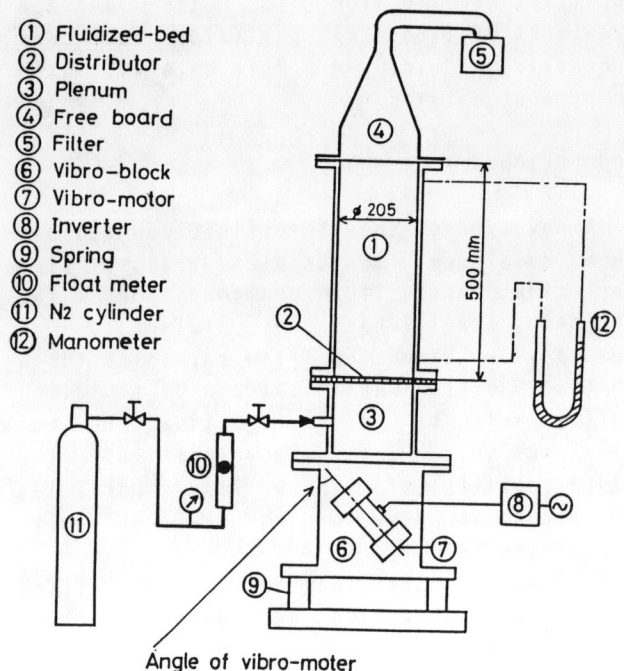

Fig. 1 Experimental apparatus

① Fluidized-bed
② Distributor
③ Plenum
④ Free board
⑤ Filter
⑥ Vibro-block
⑦ Vibro-motor
⑧ Inverter
⑨ Spring
⑩ Float meter
⑪ N₂ cylinder
⑫ Manometer

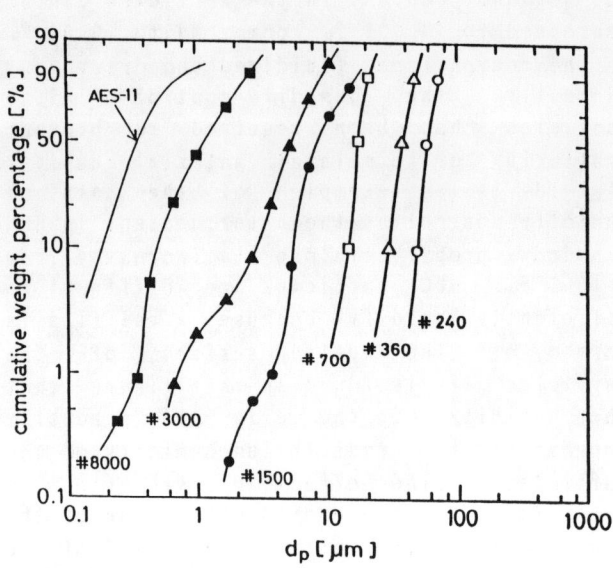

Fig. 2 Size distribution of alumina particles

Table 1 Average size and bulk density of particles

Particles		$\bar{d}_p$ [μm]	ρ_b [kg·m⁻³]
Al_2O_3	(AES-11)	0.4	1200
Al_2O_3	(#8000)	1	900
Al_2O_3	(#3000)	5	1100
Al_2O_3	(#1500)	10	1400
Al_2O_3	(#700)	24	1700
Al_2O_3	(#360)	48	2000
Al_2O_3	(#240)	80	2000
TiO_2	(A-100)	0.148	620
Activated carbon	(AC)	<20	280
zinc-oxide	(ZW)	<1	620
$MgCO_3$	(AM-50)	<10	160
Foamy polystyrene	(PS)	180	640
SiO_2	(FPS-1)	<1	72
SiO_2	(CP100)	2~3	140
$CaCO_3$	(CC)	<5	480

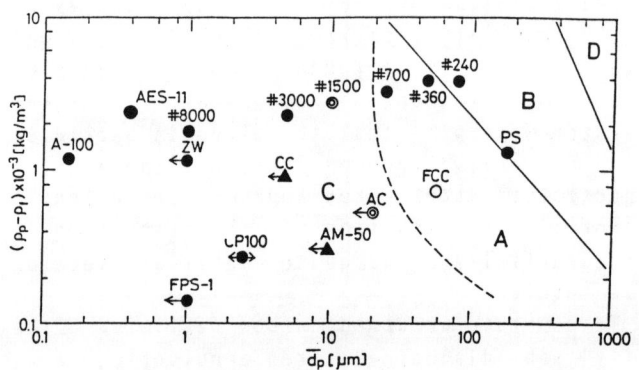

Fig. 3 Geldart's classification of particles

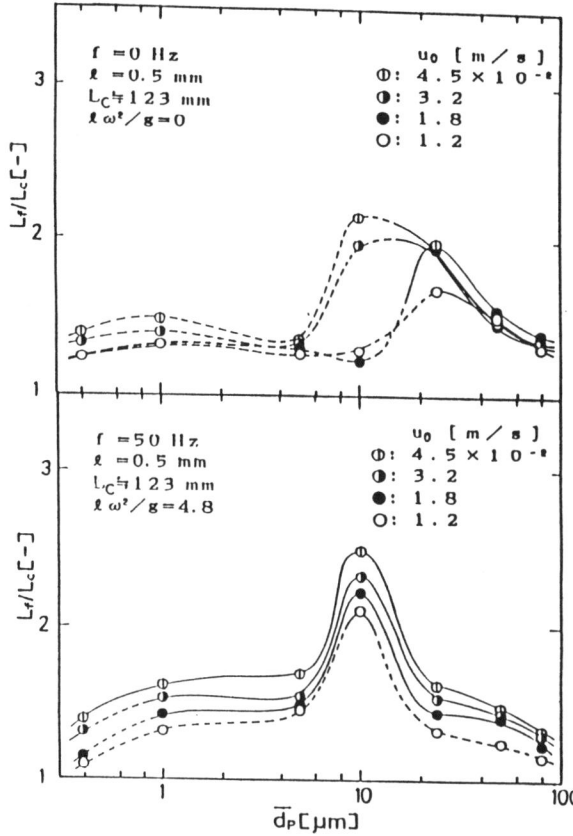

Fig. 4 Effect of vibration upon the bed expansion

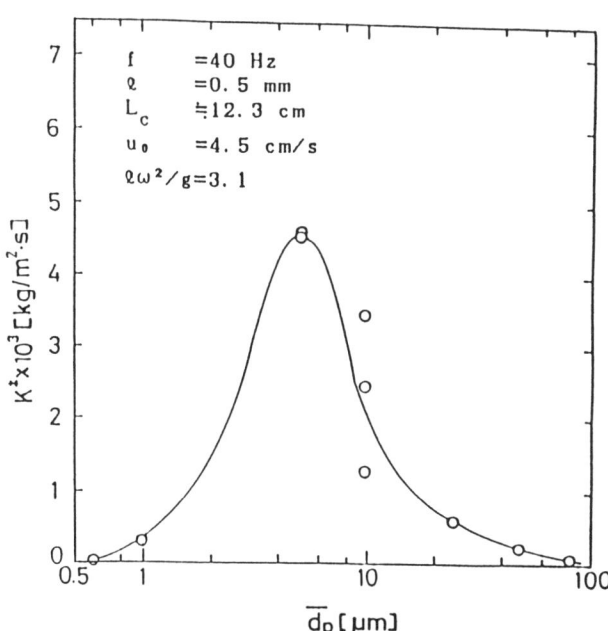

Fig. 5 Elutriation rate of alumina particles

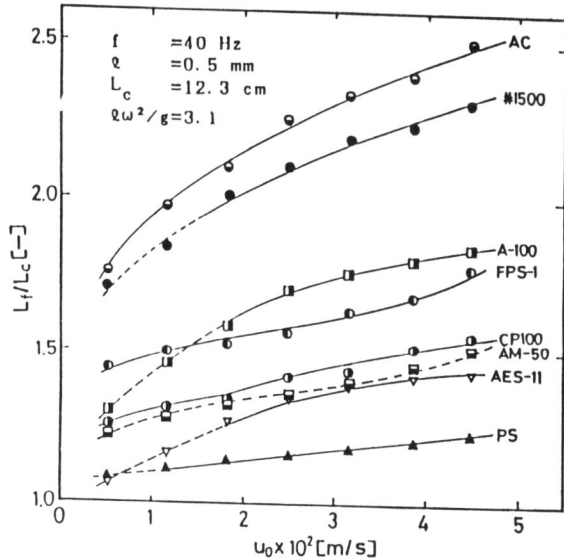

Fig. 6 Bed expansion and fluidization state of various fine particles

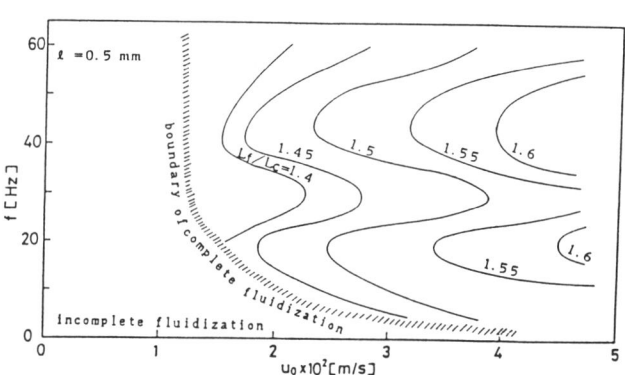

Fig. 7 Vibro-fluidization of 1 μm alumina particles

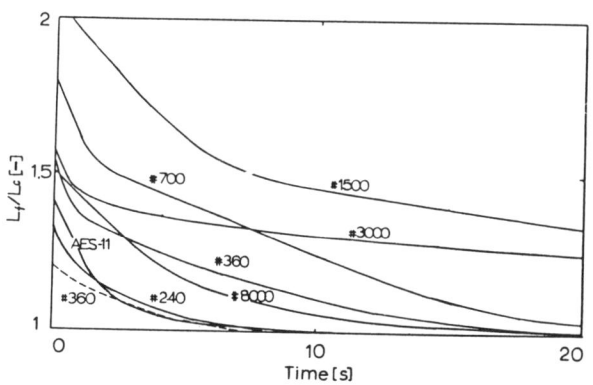

Fig. 8 Collapse of vibro-fluidized bed of alumina particles

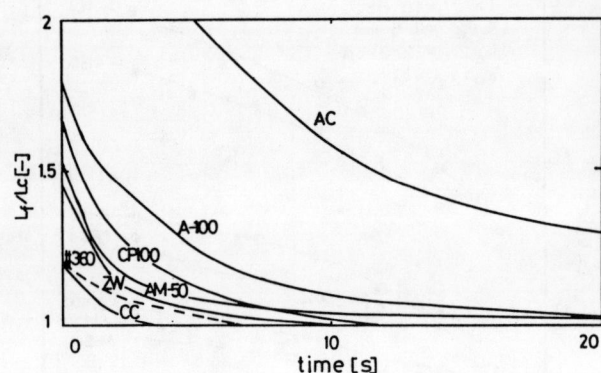

Fig. 9 Collapse of vibro-fluidized bed of various fine particles

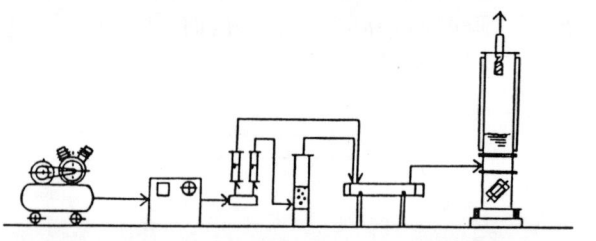

compressor, condensor, evaporater, heater, vibro-fluidized bed

Fig. 10 Vibro-fluidized bed test unit for drying and humidification

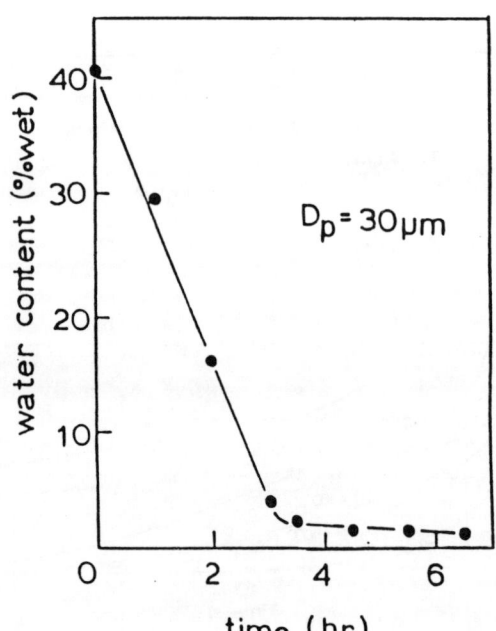

Fig. 11 Vibro-fluidized bed drying of organic fine particles

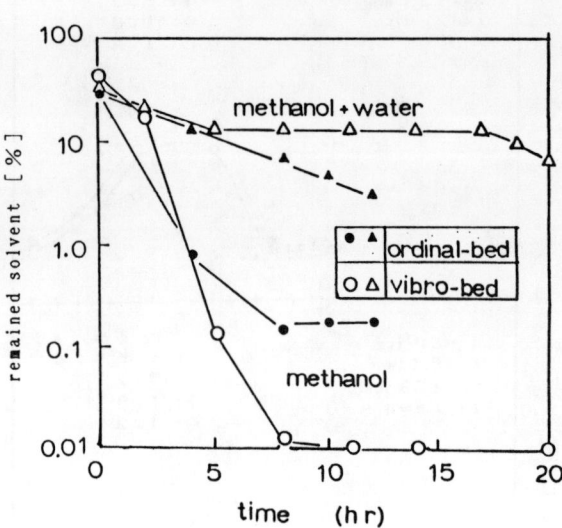

Fig. 12 Removal of methanol from organic particles

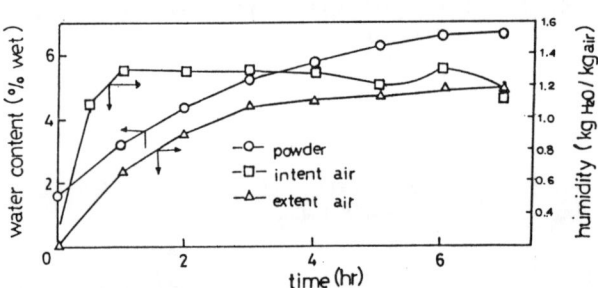

Fig. 13 Humidity control of fine HPC particles

FLUIDIZATION REGIME DELINEATION IN GAS-FLUIDIZED BEDS

S.C. Saxena, N.S. Rao and S.J. Zhou ■ Department of Chemical Engineering, The University of Illinois at Chicago, Box 4348, Chicago, IL 60680

Four different experimental procedures have been developed and adopted for probing the quality of fluidization of gas-fluidized beds and regime delineation. These involve: (i) determining the bed voidage at incipient fluidization as a function of corresponding Reynolds number (ii) measuring the local heat-transfer coefficient of a probe element as a function of fluidization number, and establishing the void-renewal frequencies from measurements of (iii) temperature and (iv) pressure histories as a function of fluidization number. These schemes have been examined in conjunction with air-fluidized beds of four different sizes of sands with the results reported here.

In a recent paper we (1) have examined the fluidization characteristics of spherical glass bead beds of four different mean sizes by utilizing three different techniques. The first scheme is based on Saxena and Ganzha (2) powder classification scheme in which the quality of bed fluidization is related to the interaction of bed particles and resulting behavior as the gas flows around the particles at incipient fluidization. As a consequence of this the bed voidage value at minimum fluidization, ϵ_{mf}, exhibits a unique relationship with Re_{mf} and is the basis of this powder classification scheme. At higher velocities, bubbles appear in the bed and the emulsion-phase behavior to some extent, and the overall gross behavior of the bed to a large extent is influenced by the bubble phase. The characteristics of the emulsion phase are still considered to be governed by the criteria developed by Saxena and Ganzha (2) in view of the two-phase theory of fluidization (3). The influence of the bubble phase on the gross behavior of a fluidized bed depends on the fluidizing velocity and on the design of the fluidized-bed system. For small particles, bubbling regime changes to bubble coalescing regime and may result into slugging regime under favorable conditions as the fluidization velocity is increased. Further, our work (1) has shown that the accurate measurement of ϵ_{mf} is difficult and its specification is further compounded by the difficulty that ϵ varies along the bed height. Additional light will be shed on this point in our present work on air-fluidized beds of different size sands.

In view of the above comments, Saxena and Rao (1) have explored two additional alternatives for characterizing the bed quality fluidization. In one scheme, the local heat-transfer coefficient is measured to an element on the heat transfer probe immersed in the bed as a function of fluidizing velocity. In the second scheme, the temperature history of this probe element is measured at various fluidizing velocities. It has been shown (1) that these two schemes lead to results which are in good agreement with each other and are uniquely dependent on the fluidization characteristics of the bed. In a recent effort, Saxena and Rao (4) have employed pressure fluctuation history records as registered by a transducer pressure probe installed on the bed wall to characterize the fluidization quality of the bed.

It is the aim of this effort to further

examine these four different schemes for establishing the quality of fluidization in relation to sand beds which in contrast with the glass beads are nonspherical in shape. The pertinent experimental details are given in the next section.

EXPERIMENTAL DETAILS AND RESULTS

The description of the experimental facility consisting of the 0.153 m square fluidized bed with its air supply system and support equipment for preparation and sizing of the bed charge etc. are given by Saxena and Rao (1). These authors (1,4) have also described the design details of the 12.7 mm diameter temperature measuring heat-transfer probe located in the fluidized bed, and that of the pressure measuring transducer probe located on the wall of the fluidized bed. A rugged stainless steel transducer appropriate for the pressure range 0-15 psig and temperature range 32 to 176° F senses the pressure signal and is energized by a regulated power supply with a variable output of 4 to 15 V d.c. The transducer assembly is accommodated in a 60 mm diameter cylindrical end cap 65 mm long and is mounted flush with the bed wall. It communicates with the bed through a 6 mm drilled channel and is equipped with a fine screen press cover at its end in contact with the bed. The latter prevents solids from entering into the channel. Both, temperature and pressure measuring probes, are located across each other at an elevation of 280 mm above the air distributor plate.

The electrical signals from the temperature and pressure probes are measured and recorded on a Hewlett-Packard data acquisition system. It consists of HP3852A data acquisition unit having a relay mux, to which the voltage signals from the probes are fed, and a voltmeter for measuring the voltage signals. The voltage signals are then sent to the HP98563 compiler where the appropriate calibration program converts them into temperature or pressure measurements. These are visually displayed on the HP98786A monitor or printed on HP2225A printer.

These results are graphically reproduced on the HP color pro-plotter.

Experiments have been conducted with 641, 1312, 2161 and 3072 μm sand beds and their size distributions are given in Table 1.

Table 1. Size distribution of sands.

U.S. Sieve number	Avg. Size (μm)	Mass Fraction of solids retained			
4-5	4375	0.0013			
5-6	3675	0.0212			
6-7	3075	0.9408	0.0098		
7-8	2580	0.0362	0.3623		
8-10	2180	0.0005	0.3700		
10-12	1850		0.1881	0.0436	
12-14	1550		0.0483	0.3640	
14-16	1290		0.0214	0.3664	
16-18	1090			0.1653	
18-20	925			0.0557	0.0042
20-30	725			0.0069	0.6810
30-35	550				0.2190
35-40	462.5				0.0743
40-45	390				0.0163
45-50	327.5				0.0051
Avg. particle diameter (μm)		3072	2161	1312	641

The bed pressure drop values for two typical sands for increasing and decreasing gas velocities are shown in Figure 1. The computed values of minimum fluidization velocities are listed in Table 2. For characterization of bed fluidization quality according to the scheme of Saxena and Ganzha (2), the knowledge of Re_{mf} is essential. For spherical particles this is easily obtained from the following relation:

$$Re_{mf} = d_p\, U_{mf}\, \rho_g/\mu_g. \qquad (1)$$

Computed values of Re_{mf} from Equation (1) using experimental U_{mf} values (Table 1) are given in Table 2A. For nonspherical particles this is a bit complicated and the general procedure which relates the Reynolds number to Archimedes number on the basis of Ergun correlation (5) at incipient fluidization is adopted here. Several efforts along this line have appeared in the literature from time to time following the pioneering work of Wen and Yu (6). We will follow here the recent result of Chen (7) who has shown that

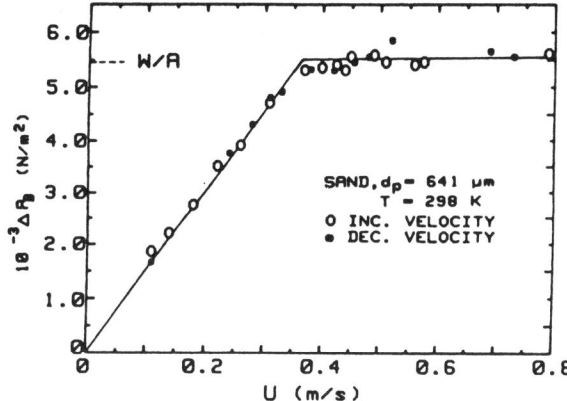

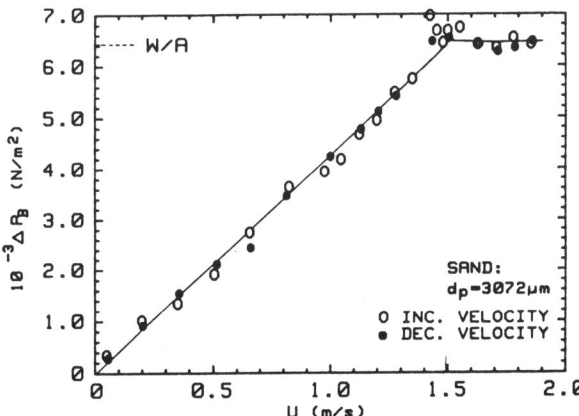

Figure 1. Variation of ΔP_B with increasing and decreasing U.

Table 2A. Properties of beds and bed particles.

Bed material	d_p (μm)	ρ_s (kg/m³)	ρ_{bulk} (kg/m³)	U_{mf} (m/s)	ϵ_{mf} (-)	Re_{mf} (-)	Ar (-)	Group
Silica Sand	3072	2628	1477	1.48	0.465	302	2.73×10⁶	III
Silica Sand	2161	2638	1610	1.28	0.425	184	9.55×10⁵	IIB
Silica Sand	1312	2533	1610	1.02	0.452	89	2.05×10⁵	IIB
Silica Sand	641	2542	1562	0.37	0.485	15.7	2.40×10⁴	IIA

Table 2B. Computed values of Re_{mf} from equation (2)

d_p (μm)	$Re_{mf}(\phi=1.0)$	$Re_{mf}(\phi=0.5)$	$Re_{mf}(\phi=0.25)$
3072	302	360	428
2161	167	201	242
1312	64	80	99
641	12	17	23

$$Re_{mf} = [(33.67\,\phi^{0.1})^2 + Ar/(24.5\,\phi^{0.45})]^{1/2} - 33.67\,\phi^{0.1}. \qquad (2)$$

ϕ may be approximated as 1.0 for near spherical particles, 0.5 for sharp particles, and 0.25 for all others. Variations to the original correlation of Ergun (5) have been proposed (8) but we will continue to use Equation (2) in this work which is based on Ergun correlation (5). The Archimedes number is computed from the following relation

$$Ar = \frac{d_p^3\, g\, \rho_g (\rho_s - \rho_g)}{\mu_g^2}, \qquad (3)$$

for use in Equation (2).

In Table 2A the values of Archimedes number obtained from the relation of Equation (3) are listed. Re_{mf} values are then obtained by using Ar values in Equation (2). These Re_{mf} values based on experimental values of U_{mf} (Table 2A) are in good agreement with the values listed in Table 2B, recalling that the sand particles are nearly spherical with a probable value of $\phi = 0.8$. ρ_s values given in Table 2A were obtained by introducing a known mass of sand in a graduated cylinder and determining the void volume filled with a known volume of water. ρ_{bulk} refers to the density of the sand bed where the bulk void volume is neglected. This is readily determined by the knowledge of the volume of a sand bed whose mass is known. The last column of Table 2A indicates the group number of Saxena and Ganzha (2) particle classification scheme to which each sand charge belongs.

The fluidized bed has also pressure taps installed in its wall at heights of 4, 21 and 36 cm above the distributor plate. Manometers of suitable density liquids are used to measure the pressure drops across the bed sections bounded between taps located at elevations of 4 and 21 cm (referred to as region 1), and those at 21 and 36 cm (referred to as region 2) as a function of fluidizing velocity. The bed voidages of these two regions 1 and 2 were then computed form the pressure drop data at each fluidizing velocity by the following relation:

$$\Delta P = L g (1 - \epsilon)(\rho_s - \rho_g). \qquad (4)$$

L is 17 and 15 cm for regions 1 and 2 respectively, and the corresponding bed voidage values are represented by ϵ_1 and ϵ_2. In Figure 2 the variation of ϵ_1 and ϵ_2 are shown as a function of U for all the four sand beds. These results clearly show that the bed voidage value (ϵ) depends on bed height above the distributor plate (h) and on the value of air fluidizing velocity (U). ϵ increases with increase in L and also with increase in U. The voidage values referring to the bed region enclosed between the heights of 4 and 36 cm is also shown in Figure 2 and ϵ_m. All the three voidage values (ϵ_1, ϵ_2 and ϵ_m) converge to a single value (ϵ_{mf}) as the air velocity is decreased to approach U_{mf}. ϵ_{mf} values determined from these plots are listed in Table 2A. However, this determination is somewhat uncertain as the upper portion of the bed is much less compacted than the lower region. In fact U is varying along the bed height due to change in pressure with elevation above the air distributor plate. We have always computed U at the mean pressure corresponding to the middle point of the bed. These difficulties in determining bed voidage and resulting uncertainties in the values of ϵ_{mf} must be kept in mind while discussing their implications in analysing the regime delineation scheme of Saxena and Ganzha (2).

Following Saxena and Rao (1), we have computed the local heat-transfer coefficient, h_1, for an element of the heated surface of the heat-transfer probe. This is obtained from the knowledge of the electrical power (Q) fed to the heater of the heat-transfer probe, and the temperature difference (ΔT) between the temperatures of the probe element and the

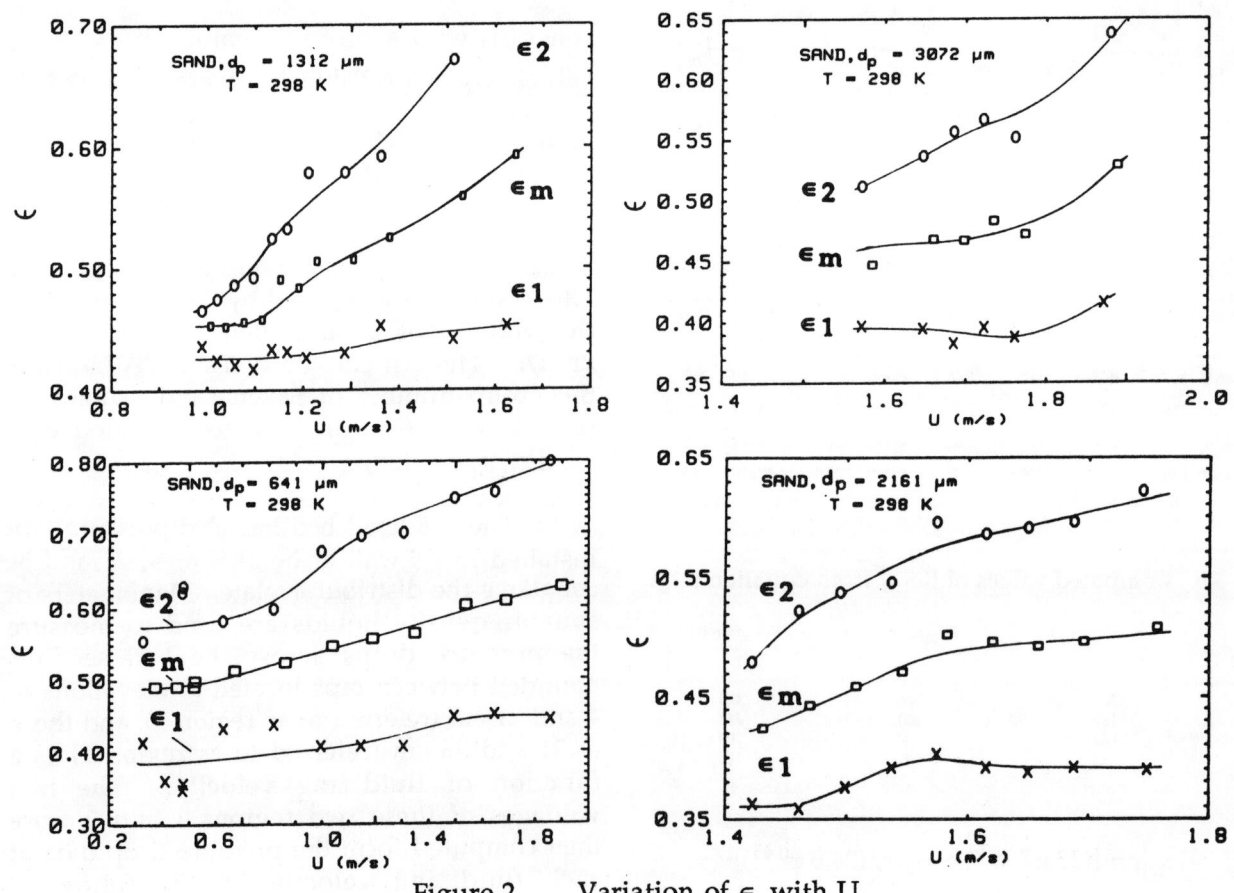

Figure 2. Variation of ϵ with U.

fluidized bed such that

$$h_1 = Q/(\Delta T) A_1. \quad (5)$$

Here A_1 is taken arbitrarily as the area of the heat-transfer probe and this choice makes h_1 values relative. This, however, is adequate for our present work where these values are used for the purpose of regime delineation only.

In Figure 3, calculated values of h_1 from Equation (4) are plotted as a function of fluidization velocity for all the four sand beds. There is a general qualitative trend of h_1 variation with increasing U implicit in these plots for all the four particles. h_1 increases with increasing U in the free-bubbling regime, and approaches a constant value with increasing U in the turbulent regime.

The temperature history records of a local heat-transfer probe element have been measured for all the four sand beds as a function of U/U_{mf} and a typical plot is displayed in Figure 4. The presence of a gas void at the heat-transfer probe element surface is reflected in its monotonic increase in temperature with time while its replacement with emulsion-phase element will make its temperature to decrease monotonically with time. Thus, the records of the type of Figure 4 referring to ten second time period may be analysed quantitatively in terms of solids or void (bubble) renewal frequency, f_{VT}. These void-frequency values are listed in Table 3. It will be seen that f_{VT} values are approximately constant over a range of U/U_{mf} values, and then exhibit another approximately constant f_{VT} value over a higher range of U/U_{mf}

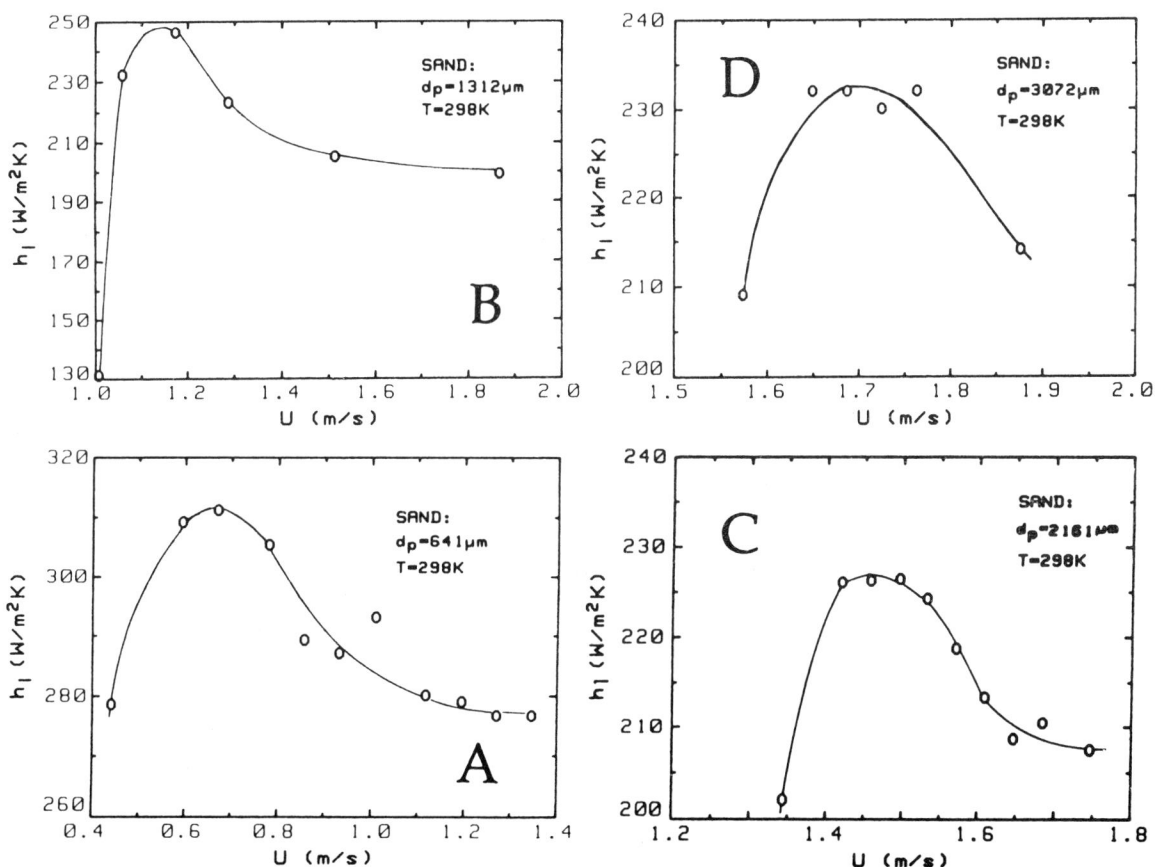

Figure 3. Variation of h_1 with U for sand beds. (A) 641, (B) 1312, (C) 2161, (D) 3072 μm.

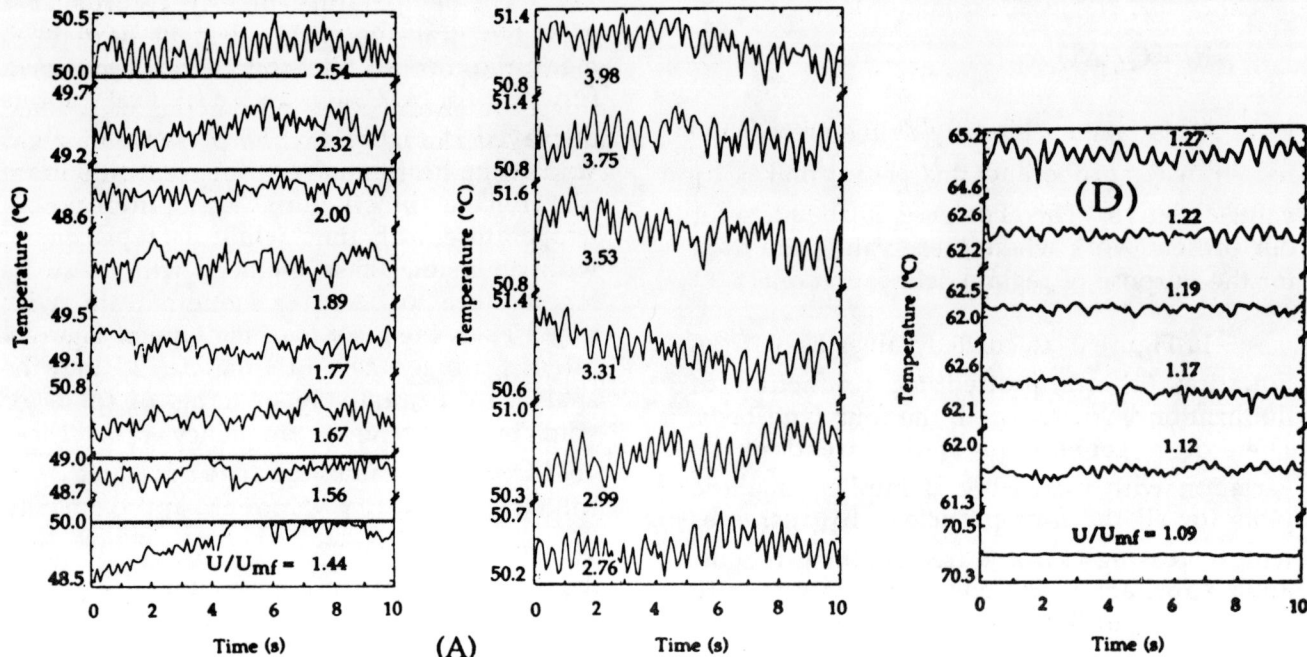

Figure 4. Temperature history of local element of heat-transfer probe at various air velocities for sands beds. (A) 641, (D) 3072 μm. The sampling rate is 100 Hz.

values. This trend is exhibited for all the four sand beds. Further the f_{VT} values corresponding to higher U/U_{mf} range is somewhat smaller than that corresponding to the lower U/U_{mf} correspond to the transition from bubbling to coalescing bubble or slugging regime.

The pressure-history plots for the four sands at different fluidizing air velocities have been taken and a typical plot is shown in Figure 5. These records were taken at ambient temperature and pressure for a period of ten seconds at a sampling rate of 100 Hz. The void-renewal frequencies, f_{VP}, determined form these plots at each air velocity are also listed in Table 3. A pressure peak in these plots represents the approach, arrival and passing away of a pressure wave at the probe surface. The pressure wave is identified with the formation and passing by of a classical bubble or a void close to the sensing probe, its propagation in the bed region above the probe and finally escape from the bed surface. The count of such pressure waves per second will thus be a measure of bubble frequencies, represented here by f_{VP}. The records of these pressure histories are more thermally stable over a longer period of time as compared to those of temperature histories recorded by the temperature measuring heat-transfer probe.

DISCUSSION OF RESULTS

The four different sized sand bed particles examined here belong to groups IIA (641 μm), IIB (1312 and 2161 μm), and III (3072 μm) of the Saxena and Ganzha (2) powder classification scheme. These are explicitly stated in Table 2A where other relevant properties of these sands and of their beds are also listed. The variation of ϵ_{mf} with Re_{mf} is not in quite accord with the expectations of Saxena and Ganzha scheme (2). For IIB powders ϵ_{mf} should increase with increase in Re_{mf}, while the listings of Table 2 exhibit an opposite trend. It appears that a larger uncertainty involved in determining ϵ_{mf} as evident from the plots of Figure 2 is mainly responsible for this discrepancy. It is also

Table 3. Record of void-renewal frequencies (f_{VP} and f_{VT}, Hz) as a function of U/U_{mf}.

d_p = 641 μm			d_p = 1312 μm		
U/U_{mf}	f_{VP}	f_{VT}	U/U_{mf}	f_{VP}	f_{VT}
1.44	2.0	2.9	1.01	1.8	2.9
1.56	1.9	3.2	1.03	2.1	-
1.67	2.3	3.0	1.07	2.1	2.5
1.77	2.3	3.2	1.11	2.1	-
1.89	2.3	3.0	1.15	1.7	2.6
2.00	2.4	3.2	1.18	1.6	-
2.32	2.1	3.4	1.22	1.2	-
2.54	1.7	3.0	1.30	1.3	2.6
2.76	1.5	3.0	1.37	1.2	-
2.99	1.5	2.7	1.52	1.4	2.1
3.31	1.9	2.9	1.64	1.3	-
3.53	1.6	2.9	1.84	-	2.2
3.75	1.6	2.6			
3.98	1.5	2.9			

			d_p = 2161 μm		
4.20	1.6		1.03	1.5	2.3
4.40	1.5		1.05	1.3	2.2
4.62	1.5		1.08	1.5	2.0

d_p = 3072 μm					
			1.11	1.1	1.9
1.09	2.3	2.7	1.13	1.1	1.9
1.12	1.9	2.5	1.16	1.0	1.8
1.17	1.4	2.3	1.19	0.9	1.8
1.19	1.0	2.4	1.27	0.9	1.7
1.22	1.1	2.0			
1.27	1.1	2.0			

possible that for nonspherical particles, the fluid-flow pattern and the resulting interparticle forces are not quite additive and they balance out each other to some extent. The final result is that ϵ_{mf} does not exhibit that characteristic variation with increasing Re_{mf} (9) as found for spherical particles (1). Lastly, the experimental procedure employed here for the measurement of ϵ_m yields a more representative value reflecting the gross behavior of the fluidized bed instead of the pure emulsion-phase value needed to accord with the classification scheme of Saxena and Ganzha (2). Probably an experimental technique reflecting more the local porosity will be appropriate for obtaining ϵ_{mf} to prove a decisive check of the predictions of Saxena and Ganzha (2) scheme. One such experimental technique is proposed by Bolens et al. (10) employing an impedance probe providing information relating to the local porosity fluctuations in the bed.

The local heat-transfer data of Figure 3 clearly represent the existence of three different regimes in all beds of sand particles except the largest where only two different fluidization regimes are encountered. It would be interesting to evolve a suitable regime delineation parameter as soon as our ongoing and planned experimental program of research in this direction is brought to completion. The transition from bubbling to bubble coalescencing regime takes place in all cases except for the smallest particle bed (641 μm) at fluidization number (U/U_{mf}) of about 1.15. The transition from slugging to turbulent regime seems to be approaching for the two smallest particles around fluidization numbers of about 3 and 1.89 respectively. For the bed of next larger size particles (2161 μm) this transition may occur at a fluidization number greater than about 1.4. These transition limits may help in analysing the f_{VP} and f_{VT} data and in establishing the equivalency of these two techniques in relation to the heat transfer based technique.

f_{VT} data given in Table 3 fairly well accord to the approximate fluidization numbers derived for regime delineation on the basis of local heat-transfer data. For 641 μm, the mean values of f_{VT} for the three regimes are 3.1 ± 0.1, 3.2 ± 0.2 and 2.9 ± 0.1. The discrimination between the free bubbling and bubble coalescing regime appears to be more ambiguous here than for the heat transfer data. For 1312 μm mean particle bed, the f_{VT} values for the first two regimes are 2.7 ± 0.2 and 2.3 ± 0.2 respectively. For 2161 μm mean particle bed, the fluidization number discriminating the first two regimes based on f_{VT} data appears to be 1.08. This is to be contrasted with the

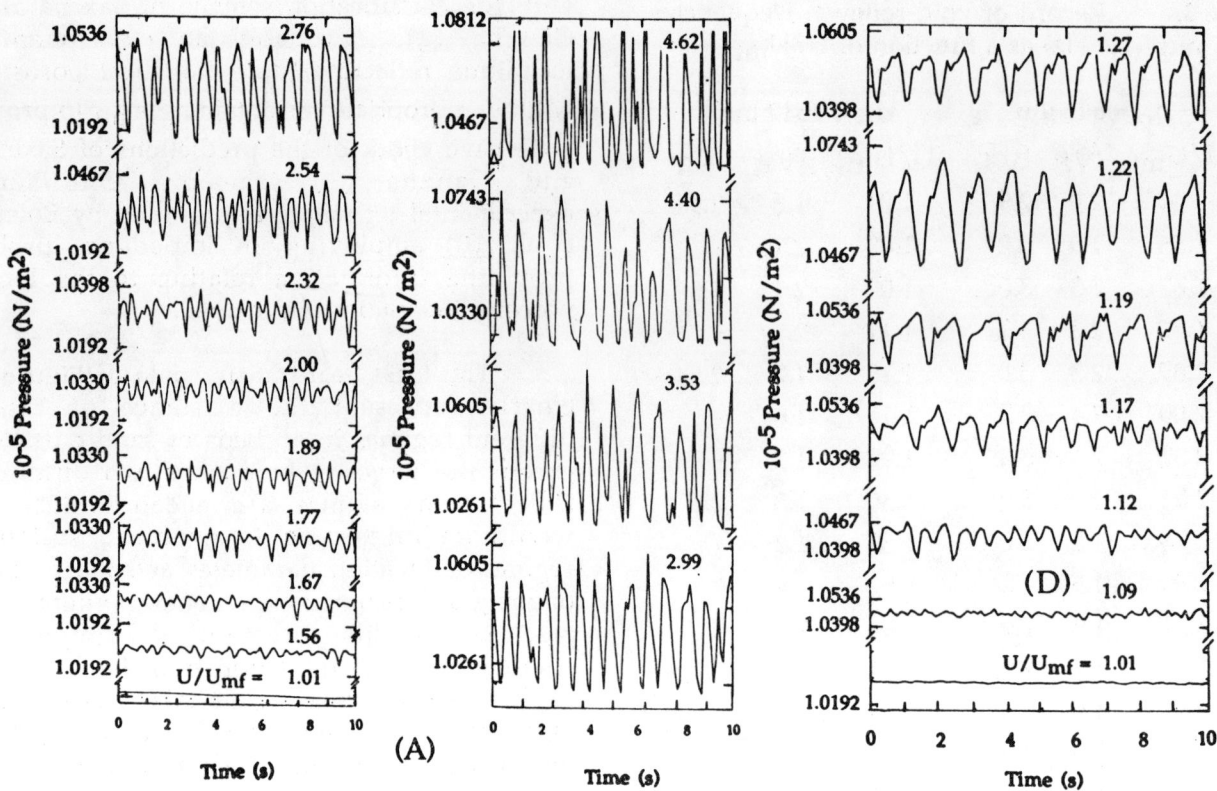

Figure 5. Pressure history of air fluidized beds of sands at a sampling rate of 100 Hz. (A) 641, (D) 3072 μm.

value of 1.15 obtained on the heat transfer data. The mean f_{VT} values for the two regimes are 2.2 ± 0.1 and 1.8 ± 0.1. Similarly for the 3072 μm mean particle diameter, the fluidization number distinguishing the two fluidization regimes based on f_{VT} is 1.19 which compares favorably with the mean value of 1.15 derived from heat transfer data. Further ongoing work will help to synthesize the independent conclusions drawn about fluidization quality from these two techniques.

f_{VP} data as reported in Table 3 are consistently smaller than the corresponding f_{VT} data for all the four sands at all fluidization numbers. This is explained and understood on the basis of data measured at different locations of these probes in the fluidized bed. The heat transfer probe is in the central region of the bed and it registers all the bubbles which are formed in the central region of the bed and those which may shift inwards from the outer region of the bed to escape through the central region of the bed. On the other hand, the pressure probe detects and responds only to those bubbles which propagate through the outer region and close to the wall. This might be the reason why f_{VP} values are consistently smaller than the corresponding f_{VT} values.

f_{VT} and f_{VP} data exhibit similar dependence on fluidization number and this may assist in establishing more dependable values of the fluidization number characteristically delineating the different fluidization regimes. However, f_{VT} values appear to be more reliable for this purpose as these reflect the fluidization behavior of the central region of the bed in contrast to f_{VP} values. A more comprehensive analysis along this line with the availability of additional gas-solid systems data will provide a more broad base to evolve a dependable regime delineation scheme. We propose to present such an analysis in the future.

In summary, the knowledge of such

parameters as ϵ_{mf}, h_1, f_{VT} and f_{VP} can assist in characterizing the quality of fluidization and hence in discrimination the different fluidization regimes. The values of these parameters vary from one region of the bed to the other but this is of no consequence in developing a regime delineation scheme. The latter can be developed reliably as long as these parameters are measured for changing fluidization regimes and regime delineation criteria are developed based on the systematic changes in the values of these parameters.

ACKNOWLEDGEMENTS

This work has been sponsored by the Illinois Department of Energy and Natural Resources and its Coal Development Board through the Center for Research on Sulfur in Coal.

NOTATION

Roman Letters

- A = Fluidized-bed cross-sectional area [m²]
- A_1 = area of the heat-transfer probe [m²]
- Ar = Archimedes number [-]
- d_p = mean particle diameter [m]
- f_{VP} = bubble or void-renewal frequency determined from pressure-history data [Hz]
- f_{VT} = bubble or void-renewal frequency determined from pressure-history data [Hz]
- g = acceleration due to gravity [m/s²]
- h = height in the bed above the distributor plate [m]
- h_1 = local heat-transfer coefficient of an element of the probe [W/m²K]
- L = separation distance between two pressure probes [m]
- Q = electrical power fed to the heater [W]
- Re_{mf} = Reynolds number at minimum fluidization [-]
- T = temperature [K]
- U = superficial gas velocity [m/s]
- U_{mf} = U at minimum fluidization [m/s]
- W = weight of the solid particles in the bed [kg]

Greek Letters

- ΔP = pressure drop [Pa]
- ΔT = temperature difference [K]
- ΔP_B = bed pressure drop [Pa]
- ϵ = fluidized-bed voidage [-]
- ϵ_1 = fluidized-bed voidage of region 1 [-]
- ϵ_2 = fluidized-bed voidage of region 2 [-]
- ϵ_m = mean fluidized-bed voidage [-]
- ϵ_{mf} = bed voidage at U_{mf} [-]
- ρ_g = density of gas [kg/m³]
- ρ_s = density of solid particles [kg/m³]
- ρ_{bulk} = bulk density of bed [kg/m³]
- μ_g = gas viscosity [Ns/m²]
- ϕ = particle sphericity [-]

LITERATURE CITED

1. S.C. Saxena and N.S. Rao, Energy, 14, 811 (1989).
2. S.C. Saxena and V.L. Ganzha, Powder Technol., 39, 199 (1984).
3. D. Kunii and O. Levenspiel, Fluidization Engineering, Krieger Publishing Company, Malabar, Florida (1984).
4. S.C. Saxena and N.S. Rao, Pressure Fluctuations in a Gas Fluidized Bed and Fluidization Quality, Energy, to be published.
5. S. Ergun, Chem. Eng. Prog., 48, 89 (1952).
6. C.Y. Wen and Y.H. Yu, AIChE J., 12, 610 (1966).
7. J.J.J. Chen, Ind. Eng. Chem. Res., 26, 633 (1987).
8. I.F. Macdonald, M.S. El-Sayed, K. Mow and F.A.L. Dullien, Ind. Eng. Chem. Fudam., 18, 199 (1979).
9. A. Mathur and S.C. Saxena, Powder Technol., 45, 287 (1986).
10. G. Bolens, F. Liefhebber and C.W.J. Van Koppen, Chem Eng. Sci., 40, 365 (1985).

THE EFFECT OF TAPER ANGLE ON THE HYDRODYNAMICS OF A TAPERED LIQUID-SOLID FLUIDIZED BED

George H. Webster and Joseph J. Perona ■ Department of Chemical Engineering, The University of Tennessee, Knoxville, TN 37996-2200

Liquid dispersion and void fraction measurements were performed for three glass bead sizes at three taper angles (0°, 0.5°, 1.5°) in order to determine the effect of particle size and taper angle on the hydrodynamics of a tapered liquid-solid fluidized bed. Approximately two times more liquid mixing occurred in the 0.5° bed and three times more occurred in the 1.5° tapered bed than occurred in the 0° (cylindrical) bed. Downward flow at the wall was characteristic of smaller size particle systems which have Reynolds numbers (based on column diameter) in the laminar flow region.

The tapered fluidized bed has in some cases found application over the cylindrical bed because its decreasing axial velocity enables the tapered bed to fluidize a wider size range of particles. Fourteen studies in the literature on tapered liquid-solid fluidized beds were noted (Webster (1)). These studies include minimum fluidization and pressure drop modelling, crystallizer and bioreactor applications, as well as void fraction and liquid mixing measurements. In particular Maruyama et al. (2) measured local void fraction profiles in tapered channels with half angles up to 12.5°. They observed at similar superficial velocities that the void fractions in the tapered bed are smaller than in the cylindrical bed. This phenomenon of the tapered bed fluidizing more densely than the cylindrical bed was also observed by Webster and Perona (3). However, we correlated the local void fraction data in a more general manner, similar to the Wen and Yu correlation (4), than did Maruyama et al., who used the form of Richardson and Zaki (5).
We note that a rationale for lower average void fractions in tapered beds than in cylindrical beds has not been addressed in the literature, and a correlation which includes the effect of angle has not yet been obtained.

In our previous study (Webster and Perona (3)) liquid dispersion was measured in a 1.5° tapered bed and compared with a cylindrical bed fluidizing 0.032 cm coal and 0.01 cm glass at Reynolds numbers (based on column diameter) of less than 1000. In all

George Webster is now at
Texaco Port Arthur
Research Laboratories
Port Arthur, TX 77641

cases liquid mixing was greater in the tapered bed. The tapered bed was characterized by fluid channelling in the center and asymmetric downward flow of particles and fluid at the walls of the column. These flow patterns were not observed in the cylindrical bed.

Our present study focuses on the effect of a reduction in taper angle and an increase in particle size on the axial dispersion and void fraction distribution. Axial dispersion coefficients were determined using the imperfect pulse method of Michelsen and Ostergaard (6) and local void fraction profiles were determined using electrical conductivity measurements.

EXPERIMENTAL EQUIPMENT

Three columns were used in this investigation with taper angles of 0°, 0.5°, and 1.5°. The cylindrical column was 5.08 cm in diameter and 120 cm in height while the 0.5° column had an entrance diameter of 5.08 cm, an exit diameter of 6.83 cm and a taper height of 98 cm. The 1.5° column had an entrance diameter of 2.54 cm, an exit diameter of 7.62 cm, and a tapered height of 107 cm. At the base of each column was flanged a cylindrical tube 40 cm (0°, 0.5°) or 60 cm (1.5°) in length and of the same diameter as the column entrance. This entrance region was packed with 0.3 cm glass beads to provide a flat velocity profile (Cairns and Prausnitz (7)). A 200 mesh wire screen was used to support the bed of particles and was placed between the column and the entrance region. Water flow rates ranged from 0.2 to 200 ml/s and were provided by a 0.5 HP pump connected to a 66 l holding tank. Ports sealed with rubber septums were located in the entrance region to provide for salt tracer injection.

Both the cylindrical and the 0.5° tapered columns were constructed of plexiglass while the 1.5° tapered column was made of glass. The vertical balancing of these columns was very important, since a slight tilt caused fluid to preferentially flow up the side of the tilted column in liquid fluidization. The three columns were balanced to the accuracy of the planar levels which were used and then checked by injecting colored dye.

Three sizes of silica glass beads were purchased from Jaygo Inc. of Mahwah, NJ in size ranges listed as 0.008 to 0.011, 0.029 to 0.044, and 0.085 to 0.123 cm. These particle size ranges were narrowed by using the cylindrical fluidized bed as a particle classifier. The narrowing procedure involved fluidizing approximately 500 g of the wide size range mixture at a void fraction of 0.85. The top 50 percent of the bed particles were removed from the column and discarded. The retained particles were refluidized and this procedure was repeated three times. This larger and more narrowly sized particle distribution was then used in the fluidization experiments. Average particle diameters were determined by a best fit between the Wen and Yu correlation and the experimental cylindrical bed void fraction data using the particle diameter as the adjustable parameter. The particle diameters were determined to be 0.0114 cm, 0.0440 cm and 0.121 cm with the average difference between the Wen and Yu correlation and experimental void fraction data ranging from 1.6 to 2.4 percent.

Particle densities were determined to be 2.524, 2.55 and 2.575 g/cm^3 for the 0.0114, 0.044 and 0.121 cm particles respectively by measuring the volume of water displaced by a known mass of particle. Jaygo Inc. listed an approximate density of 2.55 g/cm^3.

Leeds and Northrup model 7077-1-204 conductivity bridges were used in conjunction with probes in order to measure the conductivities of sodium chloride salt solutions both in the presence and absence of particles. A plastic ring-type probe was used (2.2 cm in diameter, 3.6 cm in height with a wall thickness of 0.1 cm). Two stainless steel electrodes, 1.25 cm in diameter, were epoxied into the ring and soldered to insulated wires which provided current flow and supported the probe in the fluidized bed.

EXPERIMENTAL PROCEDURE

Local void fractions were measured for each glass bead size in both the 0.5° and the 1.5° tapered fluidized beds using the conductivity probe. A 0.05 percent sodium chloride solution was pumped through the columns and conductivity was measured at various heights in the bed. The conductivity reading was recorded on a voltage output strip chart recorder and was averaged over time using a resistor-capacitor which did not affect the steady state amplitude of the signal. Conductivity measurements were converted into void fraction by a separate measurement in the cylindrical fluidized bed where the average void fraction of the bed was known. The summation of the local void fractions, ε_{OL}, according to the volume-weighted void fraction, ε_{OW}

$$\varepsilon_{OW} = \int_0^{V_T} \frac{\varepsilon_{OL} dV}{V_T} \quad (1)$$

should be equal to the overall void fraction, ε, in the tapered beds. The agreement was generally within 2 percent.

Dispersion experiments were performed according to the axial dispersion model and the imperfect pulse method of Michelsen and Ostergaard (8), (6), which requires measurements of the concentration versus time response curve to a tracer pulse at two axial points in the fluidized bed. In the Michelsen and Ostergaard method, Eq. 2 is used to determine the dispersion coefficient:

$$(\Delta\mu_s)^{-2} = \tau^{-2} + 4s/\tau \, N_{Pe} , \quad (2)$$

where τ is the plug-flow residence time, L/V, N_{Pe} is the Peclet number, VL/D, s is an arbitrary parameter in the range $0.5/\tau$ to $2/\tau$ and $\Delta\mu_s$ is the actual residence time calculated from the concentration versus time measurements at the downstream (2) and upstream (1) positions:

$$\Delta\mu_s = \frac{\int_0^\infty t \, C_2(t) \exp(-st) dt}{\int_0^\infty C_2(t) \exp(-st) dt} - \frac{\int_0^\infty t \, C_1(t) \exp(-st) dt}{\int_0^\infty C_1(t) \exp(-st) dt} . \quad (3)$$

Salt tracer solution (40 g/l) was injected into the entrance region of the columns using a syringe needle attached to a lab pump. To insure that the tracer was well mixed radially before entering the column, a 6 cm portion of the entrance region at the tracer injection point was operated as a fluidized bed by placing smaller particles in the packed bed with a screen above to restrain their expansion. Dispersion coefficients were measured at four flow rates for each particle size and taper angle. In all cases the bed height was held constant by the addition of particles when flow rate was decreased.

In measuring the conductivity versus time curves in the fluidized bed the probe was lowered through the top of the column, supported by its probe wires, and centered by small wires 5 cm above the plastic ring. The probe was placed at the bottom of the bed approximately 20 cm above the support screen and 10 tracer residence time response curves were measured. This rather laborious procedure was used because a probe at the bottom of the bed substantially effected the tracer response curve at the top of the bed. Therefore the tracer distribution curves were measured with only one probe in the bed.

Mass balances were determined as the ratio of the mass measured by the probe (the product of the flow rate and the area under the concentration versus time curve) to the mass injected. These ratios varied from 0.73 to 1.15 but in all cases the differences between upstream and downstream measurement points were less than 10 percent. This indicates that the tracer measured upstream was also being measured downstream and that there was no channelling of tracer around the probes. Furthermore the equality of upstream and downstream mass recoveries indicates good radial mixing since the probes were not exactly centered.

DISPERSION RESULTS

Dispersion coefficient results along with the plug-flow and the actual residence times are presented in Table 1. Actual residence times were obtained from the differences in the means of the distribution curves. Each coefficient in Table 1 is the result of 20 measurements (10 upstream, 10 downstream). The percent error in the dispersion coefficient was calculated as the average deviation from the mean. This error is a measure of the reproducibility of an imperfect pulse injection and measurement in the fluidized bed. The percent error ranged from 6 to 25 for the two smallest particle sizes and 16 to 47 for the 0.121 cm particles. This increased error for the larger particle size is due to the smaller residence times (5 to 12 s). It was determined for 0.0114 cm particles that the error could be reduced from 25 to 10 percent by increasing the number of measurements from 5 to 10. Ten measurement points was considered adequate given the time available and the accuracy desired.

Table 1. Dispersion Coefficients and Residence Times for 0.0114, 0.044, and 0.121 cm Glass Fluidization.

Particle size(cm)	Taper Angle(°)	Velocity (cm/s)	Void Fraction	Residence Times Ideal (s)	Actual	Dispersion Coeff. (cm^2/s)
0.0114	0	0.16	0.631	312	273	8.1
		0.31	0.734	181	180	5.6
		0.45	0.810	128	125	5.0
		0.61	0.883	119	119	3.0
	0.5	0.13	0.606	352	130	10.7
		0.24	0.704	215	124	10.7
		0.39	0.775	142	106	11.5
		0.58	0.854	110	85.6	9.3
	1.5	0.12	0.594	287	125	17.8
		0.23	0.677	171	100	11.8
		0.41	0.758	109	89	15.8
		0.65	0.847	80.1	69	16.3
0.0440	0	1.4	0.653	41.2	51.9	3.6
		2.0	0.732	32.4	36.9	3.6
		3.1	0.807	22.8	25.7	10.4
		3.9	0.884	18.5	20.2	5.2
	0.5	1.0	0.623	48.0	53.6	17.5
		1.7	0.719	32.2	32.7	12.4
		2.6	0.811	23.4	24.8	14.5
		3.5	0.890	20.2	20.7	13.7
	1.5	1.0	0.630	40.8	40.3	25.7
		1.8	0.711	26.9	26.8	23.4
		2.9	0.802	18.3	18.4	26.3
		4.0	0.877	13.6	13.5	32.9
0.121	0	4.5	0.616	11.8	12.8	3.5
		6.0	0.697	10.5	10.8	5.6
		8.1	0.780	8.6	9.0	10.6
		11.1	0.879	6.2	7.0	21.5
	0.5	3.5	0.584	12.3	11.6	5.9
		4.7	0.648	10.0	9.5	12.7
		6.3	0.727	7.1	7.9	19.7
		8.1	0.792	6.4	7.0	25.9
	1.5	2.9	0.566	9.8	9.4	9.4
		4.2	0.616	7.1	7.7	22.9
		6.0	0.696	5.8	7.1	37.9
		8.4	0.767	4.6	5.5	39.1

As can be seen by comparing the plug-flow and actual residence times in Table 1, fluid channelling does not occur in the tapered bed fluidizing 0.044 or 0.121 cm glass beads or in the cylindrical bed for all bead sizes. Significant fluid channelling does occur in the 0.5° and 1.5° tapered beds fluidizing 0.0114 cm glass beads as is shown in Figure 1. The channelling appears to be approximately equal for both beds indicating that there is no dependency on taper angle.

According to Michelsen and Ostergaard (8), (6) the linearity of Eq. 3 is proof of the applicability of the axial dispersion model. All the data in Table 1 satisfy this criteria of linearity. Since the Michelsen and Ostergaard method uses the actual residence time in the dispersion coefficient calculation, the applicability of the axial dispersion model to the channelling regions for 0.0114 cm fluidization is doubtful. However, if the additional

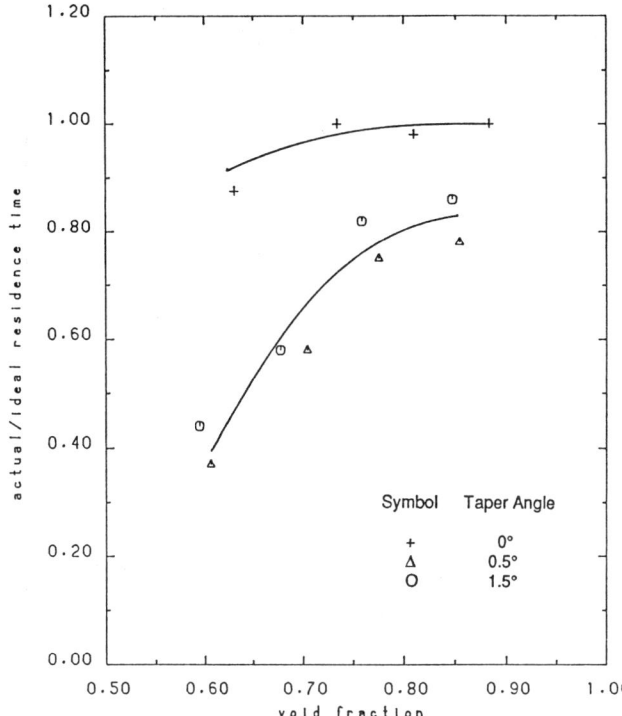

Figure 1. The ratio of actual to ideal residence time versus void fraction fluidizing 0.0114 cm glass beads.

residence time information is used to model the reactor as one of a smaller effective volume (due to channelling) the dispersion model probably can be used.

Fluidization of 0.0114 cm glass beads resulted in a symmetrical downward flow of particles at the walls of both the 0.5° and 1.5° tapered beds. This symmetric downward flow did not occur in the tapered beds for larger size glass beads or in the cylindrical bed for all size particles. A downward flow velocity of approximately 4 cm/s was observed at void fractions greater than 0.70 ($N_{Rec} > 300$) in the 0.5° tapered bed and a downward flow of approximately 5 cm/s was observed at void fractions greater than 0.76 ($N_{Rec} > 540$) in the 1.5° tapered bed. Smaller downward flow velocities (2 cm/s) were observed at void fractions less than 0.61 in the 0.5° tapered bed and at void fractions less than 0.68 in the 1.5° bed. Back flow occurs in laminar flow through diverging cones in the absence of particles at Reynolds numbers of 1375 in 0.5° and 450 in 1.5° conical vessels (Schlitching (9). It is interesting to note that the greatest downward flows occurred at large void fractions and the greatest channelling occurred at small void fractions in the tapered beds. Downward flow and channelling also appear to be independent of taper angle between 0.5° and 1.5°.

The fluidization in the cylindrical column for 0.044 and 0.121 cm glass beads was smooth but not homogeneous in that thin (approximately 10 particle diameters) disk-shaped voids were observed axially throughout the bed. These disk-shaped voids were radially uniform and appeared to travel upward through the particles in the column. This motion was also observed by Kramers et al. (10) for 0.05 and 0.1 cm glass beads. Fluidization in the 0.5° tapered bed was more violent than in the cylindrical bed. Traces of this disk-shaped void motion could be seen but unsymmetric random eddies at the wall were disruptive to this motion. There was some downward motion at the walls for larger void fractions but the velocity was less than that of the smaller 0.0114 cm particles. The 1.5° tapered bed behaved similar to the 0.5° bed fluidizing 0.044 cm glass beads, but for 0.121 cm fluidization no disk-shaped voids were observed. In all cases with the two larger size beads, the tapered fluidization was much more violent than cylindrical fluidization.

The dispersion coefficients are presented as a function of void fraction for each particle size in Figures 2, 3 and 4. Void fraction was chosen as the abscissa since according to the Wen and Yu correlation (4) it is a function of all the other fluidization variables. Dispersion coefficients increase with taper angle for all particle sizes. These dispersion results are summarized in Figure 5 as the average dispersion coefficient versus the taper angle. As shown in Figure 5 the relationship between dispersion coefficient and taper angle is almost linear for 0.121 cm and 0.044 cm glass beads, while for the 0.0114 cm beads there is a larger increase from 0° to 0.5° than from 0.5° to 1.5°. The dispersion coefficients in the 1.5° tapered bed are approximately three times as large as those in the cylindrical bed, indicating that three times more mixing occurs in the 1.5° tapered bed. The dispersion coefficients in the 0.5° tapered bed are approximately twice as large as those in the cylindrical bed. These results show that a cylindrical bed provides less axial mixing than the tapered bed and for a concentration driven reaction this means better reactor performance for the cylindrical bed. It should be noted that these results are for rather narrow particle size ranges. The advantage of the tapered bed is that it will fluidize a wide size range of particles.

The cylindrical fluidized bed dispersion data are compared to the data used by Chung and Wen (11) in Figure 6 as a Peclet number group versus the particle Reynolds number. As shown the experimental data fluidizing 0.044 cm and the 0.121 cm glass beads are within the range of data correlated by Chung and Wen. The 0.0114 cm glass bead data are at much lower Reynolds numbers than those of Chung and Wen. These dispersion

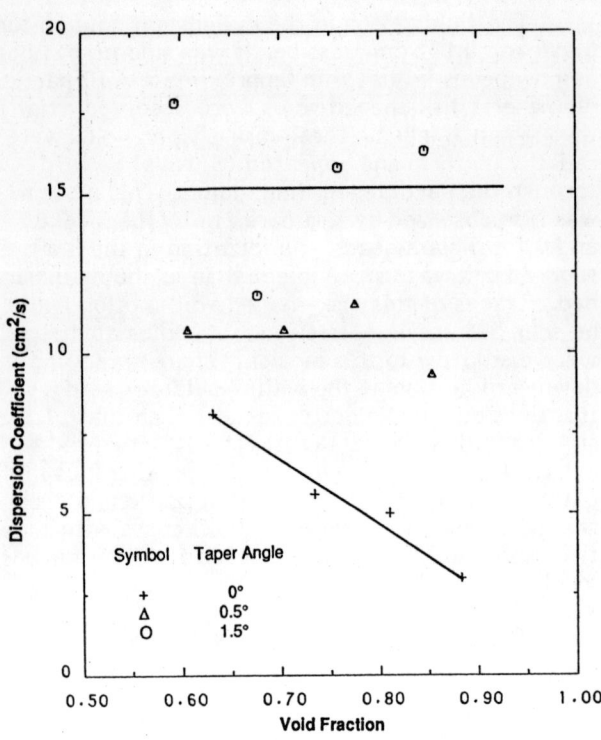

Figure 2. Dispersion coefficients for 0.0114 cm glass fluidization.

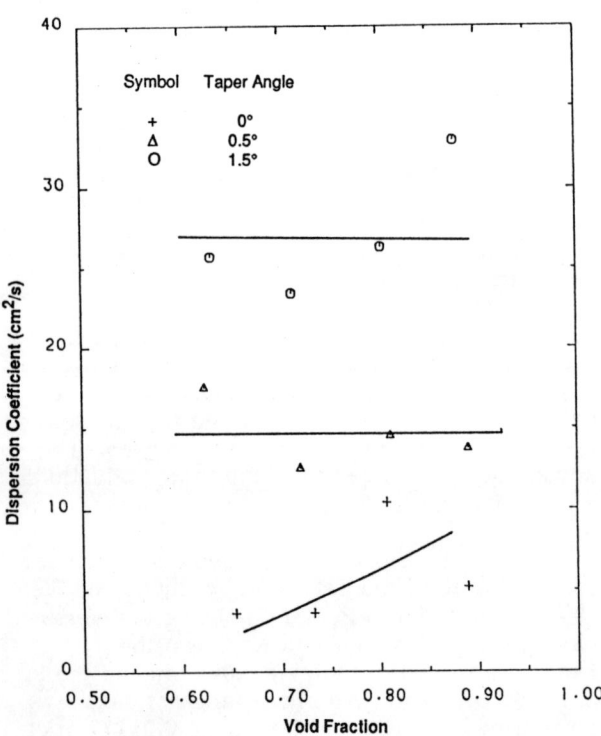

Figure 3. Dispersion coefficients for 0.044 cm glass fluidization.

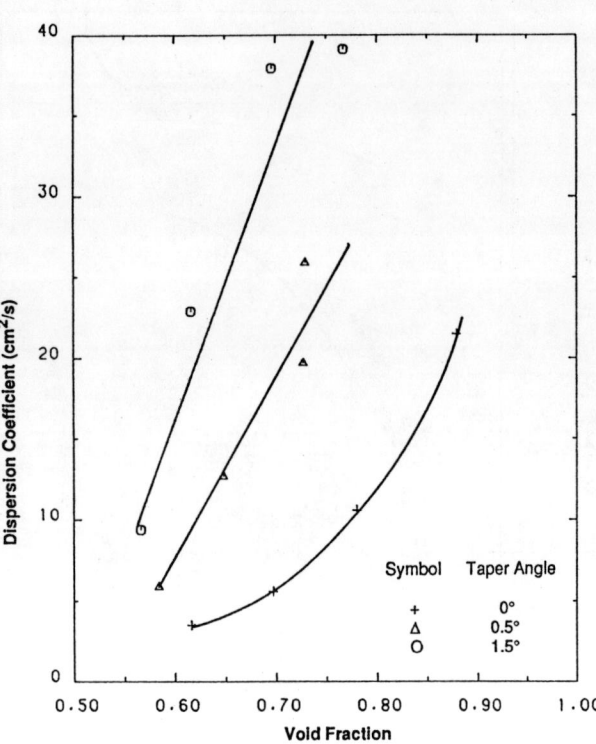

Figure 4. Dispersion coefficients for 0.121 cm glass fluidization.

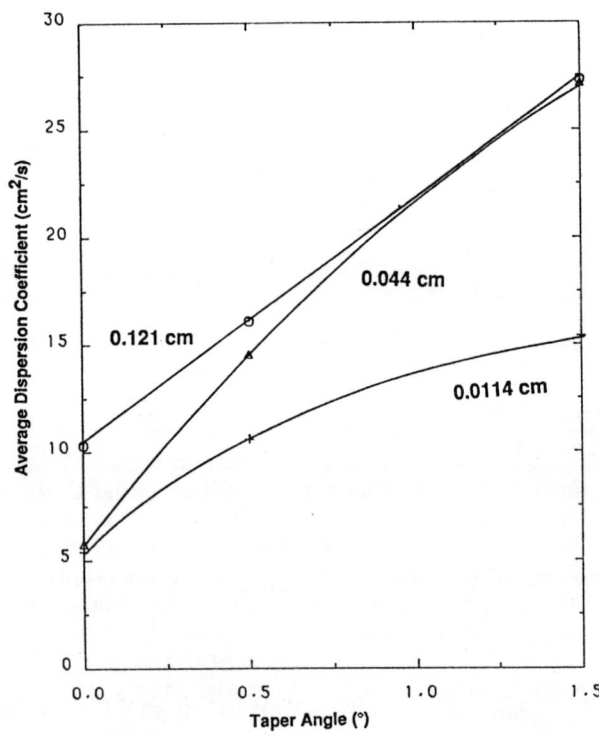

Figure 5. The average dispersion coefficient versus taper angle for glass bead fluidization.

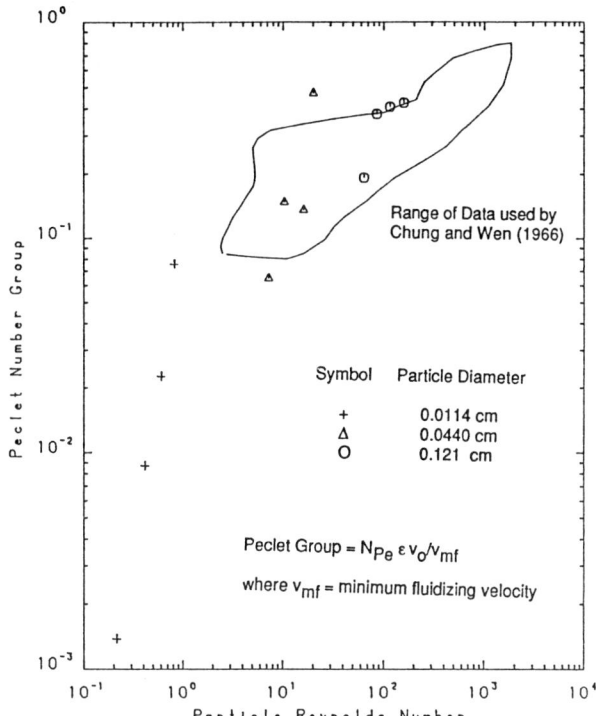

Figure 6. Comparison of experimental cylindrical bed dispersion coefficients to the data correlated by Chung and Wen (1966).

coefficients for the 0.0114 cm glass beads decrease with void fraction which is consistent with what was previously found for 0.032 cm coal (Webster and Perona (3)) and is further proof that the decrease in dispersion coefficient is a characteristic of small particle fluidization. In the tapered beds, as is shown in Figure 2 and Figure 3, the dispersion coefficients do not vary with void fraction for the 0.0114 and 0.044 cm glass beads. There is however a marked increase in the dispersion coefficient with void fraction for 0.121 cm beads in Figure 4.

The effect of the dispersion coefficient on the particle diameter is summarized in Figure 7 as the average dispersion coefficient versus the particle diameter. Although the dispersion coefficient does vary with void fraction in some cases, an average void fraction is informative since dispersion coefficients were measured at four equally spaced void fractions in each case. As shown in Figure 7 there is a substantial increase in tapered bed dispersion coefficients as the particle diameter is increased from 0.0114 to 0.044 cm, but only a slight increase from 0.044 to 0.121 cm.

LOCAL VOID FRACTION RESULTS

The local void fraction data are presented versus the local velocity in Figures 8, 9, and 10 fluidizing 0.0114, 0.044 and 0.121 cm glass beads. In Figure 9 data are presented for two particle loadings in the 0.5° tapered bed and are indistinguishable. In general the local void fraction results are similar to the overall void fraction results in that the tapered beds fluidize more densely than the cylindrical bed above a certain void fraction.

The local void fraction data in both the 0.5° and the 1.5° tapered bed were correlated in a similar way to the Wen and Yu correlation (4) for cylindrical beds. The voidage function $f(\varepsilon)$, defined as the ratio of the drag forces on a single particle in a multiparticle system to one in an isolated system, was plotted versus the void fraction for all three particle sizes. The relationships were found to be

$$f(\varepsilon) = 1.02 \, \varepsilon^{-4.47} \quad (4)$$

for the 0.5° bed and

$$f(\varepsilon) = 0.628 \, \varepsilon^{-5.22} \quad (5)$$

for the 1.5° tapered bed. The average absolute error with respect to the experimental data was 11 percent for Eq. 4 and 13 percent for Eq. 5.

Eq. 4 and Eq. 5 were substituted into the definition of the voidage function according to Wen and Yu. By solving for the void fraction the following correlations were obtained

$$\varepsilon = \left(\frac{18.4 \, N_{Re} + 2.75 \, N_{Re}^{1.687}}{N_{Ga}}\right)^{0.224}, \quad (6)$$

for the 0.5° bed and

$$\varepsilon = \left(\frac{11.3 \, N_{Re} + 1.7 \, N_{Re}^{1.687}}{M_{Ga}}\right)^{0.192} \quad (7)$$

for the 1.5° bed. Eq. 6 and Eq. 7 are compared with experimental data in Figure 8, Figure 9 and Figure 10. The best fit of the experimental data occurs for the 0.044 cm particles in Figure 9.

Eq. 6 and Eq. 7 can be used to predict void fraction for glass particle systems in 0.5° and 1.5° tapered beds and angles in between can be interpolated. A correlation which also has an angular dependence would be more appealing. In order for this to be achieved the fundamental question as to why the tapered bed fluidizes more densely than a cylindrical bed must be answered. From the results of this study it can be concluded that fluid channelling or downward flow is not a cause of the tapered bed fluidizing more densely since neither occur for 0.044 cm or 0.121 cm fluidization.

It should be noted that the effective drag force according to the Wen and Yu formulation in

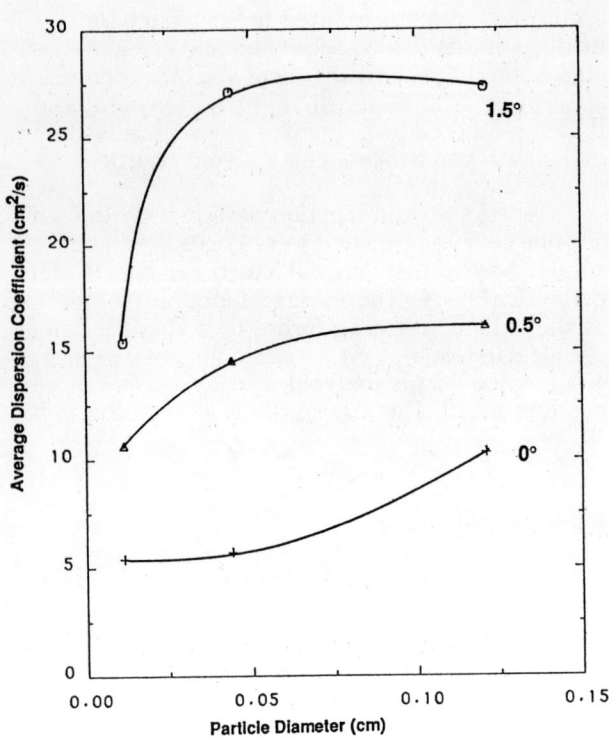

Figure 7. The average dispersion coefficient versus the particle diameter fluidizing glass beads for three taper angles.

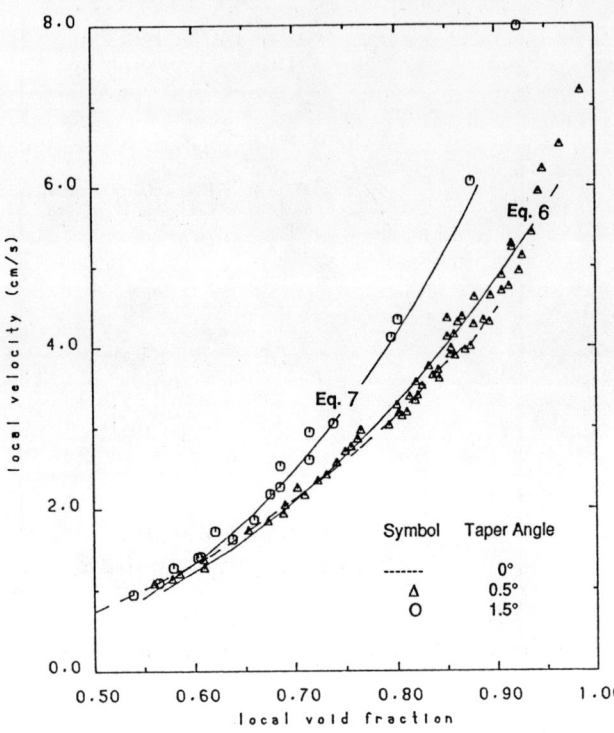

Figure 9. Local void fraction data fluidizing 0.0114 cm glass beads.

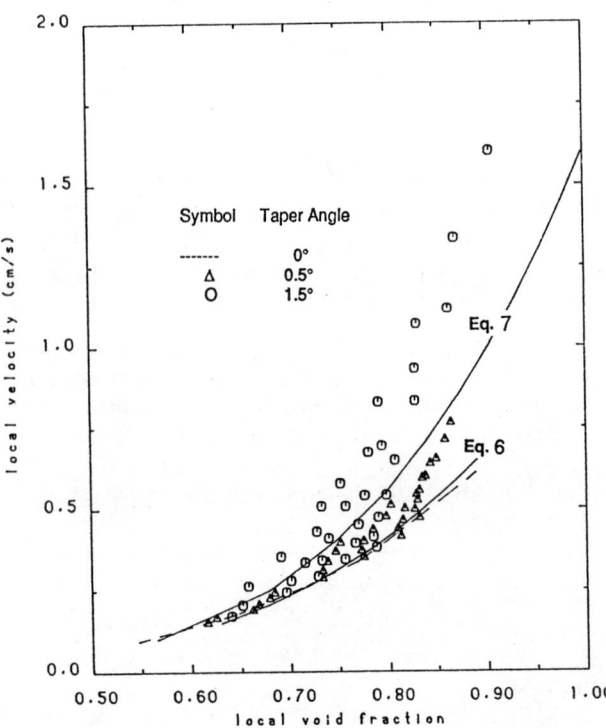

Figure 8. Local void fraction data fluidizing 0.0114 cm glass beads.

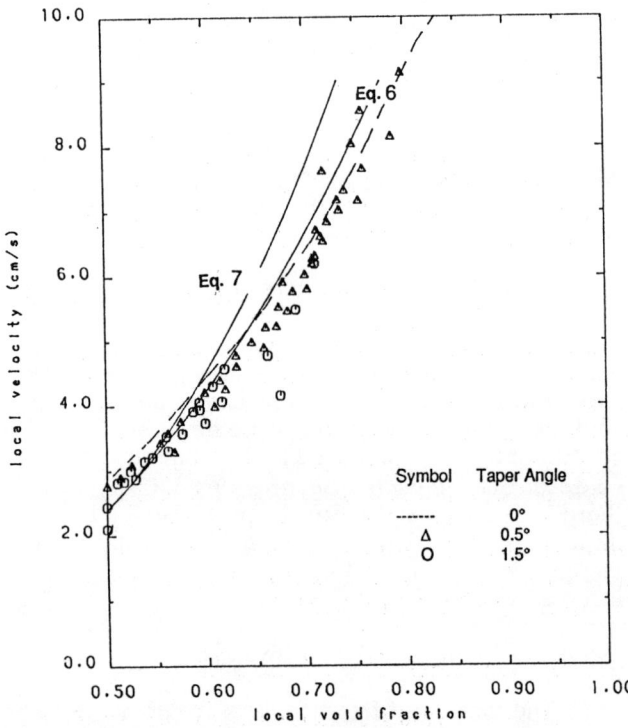

Figure 10. Local void fraction data fluidizing 0.0114 cm glass beads.

the tapered bed is less than in the cylindrical bed under similar conditions. This can be seen by comparing the voidage functions (which are proportional to the drag force) in the three beds. For a void fraction of 0.8 the voidage function in the cylindrical, 0.5° and 1.5° tapered beds are 2.85, 2.77 and 2.01 respectively. Therefore there is a significant reduction of the effective drag force on particles in the 1.5° tapered bed. This implies that the voidage function of Wen and Yu, which is based on a force balance on a single nonaccelerating particle, is not adequate for explaining this result. A more useful approach may be to explore the implications of the more general form of the particle force balance, with the mean square of the acceleration assumed to be a function of the bed geometry, but constant in time (Frazier (12)).

CONCLUSIONS

This study showed that the tapered bed's advantage over the cylindrical bed with respect to fluidizing a wide size range of particles is mitigated by increased axial mixing in the tapered bed even for small taper angles (0.5°). The downward flow and channelling which can occur in tapered beds is only a characteristic of small particle systems for which the Reynolds number is in the laminar region. For small particle systems this downward flow and channelling is independent of taper angle. It was also shown that fluid channelling is not the cause of the tapered bed fluidizing more densely than the cylindrical bed.

ACKNOWLEDGEMENT

The authors are grateful to Texaco USA for support of this research through the Texaco Fellowship in Chemical Engineering.

NOTATION

C	tracer concentration, g/l
D	dispersion coefficient, cm^2/s
d_p	particle diameter, cm
d_t	tube or column diameter, cm
f(ϵ)	voidage function defined as the ratio of the drag force on a sphere in a multiparticle system to that in an isolated one, dimensionless
g	gravitational acceleration, 980 cm/s^2
L	length between measuring points, cm
N_{Ga}	Galileo number, $d_p^3 \rho_L (\rho_s - \rho_L) g / \eta^2$, dimensionless
N_{Pe}	Peclet number, vL/D or vd_p/D, dimensionless
N_{Re}	Reynolds number, $d_p v_o / \nu$, dimensionless
N_{Rec}	Reynolds number based on tube diameter, $d_t v_o / \nu$
v	intersticial average velocity, cm/s
v_{mf}	minimum fluidization velocity, cm/s
v_o	average superficial velocity, cm/s
V_T	total fluidized volume, ml
t	time, s

Greek Letters

ϵ	experimental overall void fraction calculated as the void volume divided as the fluidized volume of bed
ϵ_{oL}	local void fraction
ϵ_{ow}	summation of local void fractions according to Eq. 1
θ	taper angle (half angle of cone), degrees
η	viscosity, g/cm·s
ρ	density, g/cm^3
ν	kinematic viscosity, cm^2/s
$\Delta\mu_s$	first moment calculated using Michelsen and Ostergaard's method (actual residence time), s
τ	residence time, either ideal (L/v) or actual ($\Delta\mu$) in the bed, s

Subscripts

1	upstream measuring position
2	downstream measuring position
G	gas phase
L	liquid phase
S	solid phase

LITERATURE CITED

1. Webster, G.H., "The Effect of Particle Diameter and Taper Angle on the Hydrodynamics of a Liquid-Solid Tapered Fluidized Bed," Ph.D. Dissertation, The University of Tennessee, Knoxville, (1988).

2. Maruyama, T., Maeda, H., and Mizushina, T., "Liquid Fluidization in Tapered Vessels," Journal Chem. Eng. Japan, 17, No. 2, 132, (1984).

3. Webster, G.H., and Perona, J.J., "Liquid Mixing in a Tapered Fluidized Bed," A.I.Ch.E. Journal, 34, No. 8, 1398, (1988).

4. Wen, C.Y., and Yu, Y.H., "Mechanics of Fluidization," Chem. Eng. Progr. Symp. Ser., 62, 100, (1966).

5. Richardson, J.F., and Zaki, W.W., "Sedimentation and Fluidization: Part I," Trans. Inst. Chem. Engrs., 32, 35, (1954).

6. Michelson, M.L., and Ostergaard, K., "Holdup and Fluid Mixing in Gas-Liquid Fluidized Beds," Chem. Eng. Journal, 1, 37, (1970).

7. Cairns, E.J., and Prausnitz, J.M., "Velocity Profiles in Packed and Fluidized Beds," Ind. Eng. Chem., 51, 1441, (1959).

8. Michelson, M.L., and Ostergaard, K., "On the Use of the Imperfect Pulse Tracer Method for Determination of Holdup and Axial Mixing," Canadian Journal Chem. Eng., 47, 107, (1969).

9. Schlichting, H., "Boundary Layer Theory," p. 81, 1st English Ed., McGraw Hill, New York City, (1955).

10. Kramers, H., Westermann, M.D., de Groot, J.H., and Dupond, F.A.A., "The Longitudinal Dispersion of a Liquid in a Fluidized Bed," Third Congr. European Fed. of Chem. Eng., Olympia, London, (June 1962).

11. Chung, S.F., and Wen, C.Y., "Dispersion of Liquid Flowing Through Fixed and Fluidized Beds," A.I.Ch.E. Journal, 14, 857, (1968).

12. Frazier, G.C., personal communication, The University of Tennessee, Knoxville (1988).

BIOLOGICAL PHENOL DEGRADATION IN A COUNTERCURRENT THREE-PHASE FLUIDIZED BED USING A NOVEL CELL IMMOBILIZATION TECHNIQUE

Kuo-Ying Amanda Wu and Keith D. Wisecarver ■ Department of Chemical Engineering, The University of Tulsa, Tulsa, OK 74104

INTRODUCTION

Fluidized bed bioreactors have, in recent years, become an important new technology for both wastewater treatment and fermentation processes. They have been shown to be effective for both aerobic and anaerobic treatment of domestic and industrial wastewaters to remove organic carbon and nitrogen (1). Fluidized bed bioreactors have a number of advantages over conventional bioreactors, including: very high holdup of biomass; prevention of washout of the microbes; lack of clogging of the biomass; increased resistance to variations in substrate concentration, toxic chemicals, heavy metals, etc.; ease of separation of cells from product stream; and large liquid-solid contacting area and high mass transfer rates (2,3,4).

In the fluidized bed bioreactor, microorganisms are immobilized onto solid support particles, which are then fluidized by the liquid stream. Cell immobilization for mixed cultures of organisms is often accomplished by the formation of a biofilm on the particle surface due to the natural excretion of adhesive exopolymers from the microbial cells. For pure cultrues, entrapment of cells within a polymeric matrix is a common technique for cell immobilization. Materials such as calcium alginate or kappa-carrageenan are commonly used for cell immobilization. However, there are problems with the physical durability of these materials. A need exists for cell immobilization supports which are durable enough to withstand long periods of operation under the shearing forces which are encountered in fluidized beds.

For aerobic biological reactions oxygen must be supplied to the mircoorganisms. If a liquid-solid fluidized bed is used, this usually requires the addition of an aeration device (such as a bubble column) to increase the dissolved oxygen level of the incoming liquid. Alternatively, an air stream can be bubbled directly into the fluidized bed creating a gas-liquid-solid, or three-phase, fluidized bed. Both two-phase and three-phase fluidized beds have been successfully applied to the aerobic biological treatment of wastewaters (5,6,7). The gas-liquid-solid fluidized bed has an inherent advantage over the liquid-solid fluidized bed due to the improved supply of oxygen to the micro-organisms and the elimination of the need for pre-aeration of the liquid. There are several disadvantages, however, to the gas-liquid-solid fluidized bed. For very small or low-density particles such as are commonly encountered in immobilized-cell bioreactors, the solids holdup decreases rapidly and particle elutriation becomes severe as the gas or liquid velocity is increased. Also, the increased agitation in the bioreactor caused by the bubble motion greatly increases the particle-particle contact and creates a significant shear force, which can result in shedding of the attached biofilm or destruction of the polymeric matrix of the particles. The gas and liquid velocities in a gas-liquid-solid fluidized bed bioreactor must therefore be carefully controlled to maintain adequate bed expansion and gas-liquid mass transfer while minimizing shear effects and particle elutriation (8).

A novel method of increasing the solids holdup and decreasing particle elutriation is the use of an inverse or countercurrent three-phase fluidized bed, in which the liquid flows downward countercurrent to the gas flow. This countercurrent flow of gas and liquid eliminates one of the chief disadvantages of the conventional (cocurrent flow) three-phase fluidized bed, namely that of elutriation of particles due to the passage of bubbles through the bed. Such a reactor configuration has been studied for particles which have a lower density than the liquid (9,10,11). However, no study has yet appeared in the literature concerning countercurrent three-phase fluidized beds in which the particle density is larger than the liquid density.

In this work, the behavior of a countercurrent three-phase fluidized bed bioreactor for phenol degradation was studied. A pure culture of a phenol-degrading organism was used, which was immobilized by a novel cell immobilization technique. Hydrodynamic studies were performed to determine the range of gas and liquid velocities over which the countercurrent three-phase fluidized bed can be operated.

MATERIALS AND METHODS

Countercurrent Three-Phase Fluidized Bed

A schematic diagram of the three-phase fluidized bed reactor is shown in Figure 1. The fluidized bed reactor is a 7.62 cm inside diameter and 2.44 m long Plexiglas column and consists of three sections. The bottom and the top sections are 0.61 m (24 inches) long each. The middle section is 1.22 m (48 inches) long. The liquid was pumped from the feed tank to the top of the reactor passing through the distributor which has 54 holes 1.59 mm (1/16 inch) in diameter. The air is fed to the distributor by two tubes from opposite sides for better gas distribution. The liquid passes through the distributor through nineteen 6.35 mm (1/4 inch) ID tubes. A stainless steel mesh, with 2 mm openings, is located at the bottom of the column just above the gas distributor. Liquid exiting from the bottom of the column is sent to a weir before going to the drain. The height of the weir can be adjusted in order to set the desired liquid level in the column.

Cell Immobilization Technique

The microorganism used in this work was a pure strain of phenol-degrading Pseudomonas #1101 obtained from Microbe Masters Inc. of Baton Rouge, Louisiana. The composition of the medium used both for growing the organisms and in the fluidized bed experiments is given in Table 1. Approximately 420 cm^3 of beads containing the immobilized cells were made by adding centrifuged cells to a mixed solution of 13.2% (w/v) poly(vinyl alcohol) (PVA) (100% hydrolyzed, average M.W. 77,000-79,000 from Aldrich Chemical Company) and 2% (w/v) sodium alginate (sodium salt, low viscosity, approx. 250 cps from Sigma Chemical Company). A small amount of medium was added to supply nutrients for the microbial cells during the solidification process. Beads were formed by cross-linking the PVA with boric acid: drops of the PVA-alginate solution were allowed to fall into ten liters of solution consisting of saturated boric acid and 1 g/L $CaCl_2-2H_2O$. The beads were kept stirred in this solution for 24 hours to complete the solidification. The final composition of the beads was 10.4% PVA and 0.02% alginate, with a cell density of 5×10^9 cells per cm^3 of bead. The alginate was added to improve the surface properties of the beads. It was found that pure PVA formed beads which had a strong tendency to clump together, while the addition of 0.02% alginate to the beads was sufficient to prevent this clumping. The diameter of the PVA-alginate beads was approximately 3 mm.

Phenol Analysis and Stripping

The phenol concentrations of all the samples in this work were analyzed by gas chromatography (Hewlett Packard model 5890A) with a flame ionization detector. The retention time for phenol was 7 minutes. The phenol concentrations in the centrifuged medium were determined by comparing the peak area to those of standard solutions.

Stripping the system of phenol was done to test for possible evaporation of phenol due to the aeration of the fluidized bed. A two liter flask with 500 mL of pH=10 NaOH solution (chosen because $pK_a=10$ for phenol) was sealed with a rubber stopper containing two glass tubes. One tube was used for releasing the gas in the flask. The other tube was immersed in the NaOH solution, while the other end of this tube was connected to the gas outlet at the top of the fluidized bed bioreactor. The top of the reactor was closely sealed by a rubber stopper so that the outlet gas stream from the reactor would go to the stripping apparatus.

Measurements of Minimum Fluidization Velocities

The PVA-immobilized bioparticles were successfully fluidized in the countercurrent three-phase fluidized bed, even though the particle density (approximately 1002 kg/m^3) was larger than that of the medium. Experiments were therefore undertaken to determine the range of gas and liquid flow rates over which the countercurrent three-phase fluidized bed bioreactor could be operated in the fluidized state. Minimum fluidization conditions for the bed under typical bioreactor operating conditions were measured in experiments using calcium alginate beads, with air and water as the fluidizing media. 600 cm^3 of 2.9 mm diameter calcium alginate beads, with a particle density of 1029 kg/m^3, were placed in the bed. The liquid level in the column was controlled by adjusting the weir height. Minimum fluidization was determined by visual observation.

The minimum gas velocity required to suspend the particles at zero liquid flow rate was first determined. The water flow was turned off and the liquid height in the column was set by adjusting the weir level. The air velocity required to fluidize the particles was determined by decreasing the air flow rate when the particles were in the suspended state until particle fluidization no longer occurred. Minimum fluidization was considered to be the point at which particles at the very bottom of the bed ceased to move independently of the neighboring particles. The liquid height was then adjusted by changing the weir height, and a new gas minimum fluidization velocity was found. The liquid heights in the column were observed to be the same as the weir levels consistently during this phase.

The gas velocity required to fluidize the particles at non-zero liquid velocity was then measured. The weir level was fixed in order to keep a stable liquid height in the column. Minimum fluidization velocities of air were measured at various liquid flow rates for each weir setting. This procedure was then repeated for other weir settings. It was observed that the height of liquid in the column tended to increase even though the weir position was kept constant when the liquid flow rates or the air flow rates were increased. The difference in heights between the liquid in the column and the weir increased as the weir level was lowered.

RESULTS AND DISCUSSION

Biological Phenol Degradation

During the phenol biodegradation experiments, air was delivered to the countercurrent three-phase fluidized bed at a rate of 1.4 L/min and the medium flow rate was set at 2.5 L/h. Table 2 shows the experimental conditions and results including the flow rates of liquid and air during the process and the change in phenol concentration. The liquid flow rate of the medium in the reactor was started at 1 L/h to allow the cells to adjust to the new medium and progress from the lag phase to the exponential phase (12). Steady state was considered to be achieved when the outlet phenol concentration remained constant, i.e. zero phenol concentration in this experiment. Usually two residence times (6.4 hours in this work) would be enough to achieve a steady state. It is suggested that the increment of the phenol concentration should not be more than 100 mg/L each time after the steady state is achieved. Complete degradation of the phenol was accomplished for inlet phenol concentrations as high as 1300 mg/L, as shown in Table 2. The phenol concentration was then adjusted suddenly to 2000 mg/L. At this point the bacteria stopped degrading the phenol and were presumed dead. This is most likely due to the accumulation of phenol concentration in the fluidized bed bioreactor, but could also be due to oxygen limitation at 2000 mg/L phenol concentration.

Foaming in the fluidized bed experiments, caused by either air flow or PVA, was a serious problem. Addition of Silicones AF93 (silicone antifoam emulsion from Southwest Silicone Company) eliminated the foaming, but at somewhat higher gas rates, an antifoam agent has the further effect of inhibiting bubble coalescence and offsetting the tendency for bubble size to increase with increasing gas holdup (1).

There are potentially three removal mechanisms for phenolic compounds: air stripping, adsorption, and biodegradation (13). It was determined that the percentage of phenol degradation due to stripping was 0.1% of the total phenol degradation. Bioadsorption was found to be small in comparison with biodegradation by a number of investigators (14,15,16). It was consequently concluded that phenol removal in the bioreactor was due almost entirely to biodegradation.

The bioparticles made from the PVA-alginate immobilization technique were extremely strong and durable, showing no sign of breakage or disintegration after several weeks of operation in the fluidized bed.

Minimum Fluidization Velocities

The overall solids holdup was determined in the experiments by visual observation of the bed height. At low overall solids holdups (i.e. high liquid levels in the column), there was a gradual change in local solids holdups from a relatively dense solids region at the bottom of the column to a relatively dilute solids region at the top. As the liquid level in the column was lowered and the overall solids holdup correspondingly increased, this distribution of solids was found to gradually disappear until, at the highest overall solids holdups, the solids distribution in the column was essentially uniform.

It was found that fluidization could be maintained in the system even at solids holdups as high as 0.55, which is close to the solids holdup at the minimum fluidization state for liquid-solid fluidized beds of 0.59 as reported by Wen and Yu (17). Minimum fluidization conditions were determined for overall solids holdups ranging from 0.08 to 0.55, and for liquid superficial velocities as high as 0.83 cm/s. Fluidization could undoubtedly be maintained at even higher liquid velocities, but the liquid level became difficult to control at higher liquid velocities because the liquid height became appreciably higher than the weir height.

The gas velocity required to fluidize the calcium alginate particles at zero liquid velocity is plotted as a function of the solids holdup in Figure 2. It can be seen that, at low solids holdup, the gas velocity required to suspend the particles approaches a limiting value, which corresponds theoretically to the gas velocity required to suspend a single particle. As the solids holdup approaches the packed bed state, it appears that the gas velocity required to suspend the particles also approaches a limiting value.

Figure 3 shows the gas velocity required to fluidize the particles at constant liquid velocity as a function of the overall solids holdup. At a superficial liquid velocity of 0.069 cm/s, the data look very similar in trend to the data for zero liquid velocity shown in Figure 2, with a monotonic increase in the required gas velocity as the solids holdup increases. At higher liquid flow rates, however, there is a clear trend for the gas velocity at minimum fluidization to decrease and then increase as solids holdup is increased. This can possibly be due to a change in the flow regime of the bed. At low liquid velicities the bed is in the coalesced bubble regime, wherein the bubbles tend to coalesce into relatively large bubbles. The presence of small and/or light particles is known to increase the tendency for bubble coalescence. The coalesced bubbles tend to result in the gas bypassing the emulsion phase rather than contributing to the fluidization of the particles, so the increased bubble coalescence at high solids holdup requires a higher gas velocity to fluidize the particles. As the liquid velocity is increased, however, the hydrodynamic force of this downward flow should help to break up the bubbles. Apparently, there is a tendency for bubble breakage and dispersion with increasing solids holdup for solids holdup below about 0.18, and for bubble coalescence with increasing solids holdup for solid holdups above this value.

The gas and liquid velocities at minimum fluidization conditions are plotted in Figure 4. It can be seen that, at low solids holdups, the gas velocity required to fluidize the particles increases with increasing liquid velocity. This is to be expected, since the downward movement of liquid should tend to inhibit particle fluidization. However, at higher solids holdups, it can be seen that the gas velocity required to suspend the particles becomes considerably greater at zero liquid velocity than at non-zero liquid velocity. It is hypothesized that the flow of liquid through the emulsion phase in the region of the distributor tends to cause some particle movement in the distributor region, allowing particles to move closer to the bubble formation region where there can be suspended by the bubble flow. This effect becomes increasingly important at high solids holdups. It can also be seen from Figure 4 that, at higher superficial liquid velocities, the gas velocity for minimum fluidization increases slightly with solids holdups.

CONCLUSIONS

The cell immobilization technique, in which a pure culture of Pseudomonas was immobilized in a poly(vinyl alcohol) gel cross-linked with boric acid with the addition of a small amount of alginate, was found to produce beads with good strength and durability with little loss of biological activity. In phenol biodegradation experiments in the countercurrent three-phase fluidized bed, nearly complete biodegradation was achieved for inlet phenol concentrations as high as 1300 mg/L.

The range of gas and liquid velocities over which the countercurrent three-phase fluidized bed could be operated were measured. It was determined that the bed could be maintained in a fluidized state at solids holdups as high as 0.55, over a wide range of gas and liquid velocities, using calcium alginate particles. The countercurrent three-phase fluidized bed would appear to have a significant potential for three-phase reactor applications due to the high solids holdup attainable.

LITERATURE CITED

1. Cooper, P., B. Atkinson, *Biological Fluidized Bed Treatment of Water and Wastewater*, E. Harwood Ltd., London (1981).

2. Rosevear, A., *J. Chem. Tech. Biotechnol.*, 34, 127 (1984).

3. Donaldson, T.L., G.W. Strandberg, J.D. Hewitt, G.S. Shields, M.R. Worden, *Environ. Prog.*, 6(4), 205 (1987).

4. Kuu, W.Y., J.A. Polack, *Biotechnol. Bioeng.*, 25, 1995 (1983).

5. Holladay, D.W., C.W. Hancher, C.D. Scott, D.D. Chilcote, *J. Water Pollut. Control Fed.*, 50, 2573 (1978).

6. Lee, D.D., C.D. Scott, C.W. Hancher, *J. Water Pollut. Control Fed.*, 51, 974 (1979).

7. Wisecarver, K., L.S. Fan, *Biotechnol. Bioeng.*, 33, 1029 (1989).

8. Fan L.S., B.E. Kreischer, K. Tsuchiya, paper 28g, presented at the AIChE Annual Meeting, Miami Beach, Fla., Nov. 2-7, 1986.

9. Shimodaira, C., Y. Yushina, *Proc. 3rd Pac. Chem. Eng. Cong.*, 4, 237 (1983).

10. Nikolov, L., D. Karamanev, *Can. J. Chem. Eng.*, 65, 214 (1987).

11. Fan, L.S., K. Muroyama, S.H. Chern, *Chem. Eng. J.*, 24, 143 (1982).

12. Zobell, C.E., *Scripps Institution of Oceanography, New Series No. 295, University of California, La Jolla*, 10, 1 (1946).

13. Lewandowski, B., S. Salerno, N. McMullen, L. Gneiding, D. Adamowitz, *Environ. Prog.*, 5(3), 212 (1986).

14. Gaudy, A.F., D.F. Kincannon, T.S. Manickam, EPA-600/2-82-075 (June, 1982).

15. Kincannon, D.F., E.L. Stover, V. Nichols, D. Medlay, *J. Water Pollut. Control Fed.*, 55, 157 (1983).

16. Petrasek, A.C., *J. Water Pollut. Control Fed.*, 55, 1286 (1983).

17. Wen, C.Y., Y.H. Yu, *Chem. Eng. Prog. Symp. Ser.*, 62, No. 62, 100 (1966).

Table 1: Composition of Stock Nutrient Medium

Composition	Per Liter
Phenol	variable
NH_4Cl	2.13 g
$CaCl_2 \cdot 2H_2O$	1.00 g
$MgSO_4 \cdot 7H_2O$	0.6 g
$MnSO_4 \cdot H_2O$	0.02 g
$FeSO_4 \cdot 7H_2O$	0.02 g
Spring Water (for trace elements)	130 mL
K_2HPO_4/KH_2PO_4 buffer, pH = 7.6-9.0	15 mL

Table 2: Results of Phenol Degradation in a Countercurrent Three-Phase Fluidized-Bed Bioreactor

FLOW RATE (L/h)	INLET CONC. (mg/L)	OUTLET CONC. (mg/L)
1.0	250	0
2.5	250	0
2.5	350	0
2.5	450	0
2.5	550	0
2.5	650	0
2.5	750	0
2.5	850	0
2.5	950	0
2.5	1000	0
2.5	1100	0
2.5	1200	0
2.5	1300	0
2.5	2000	2000

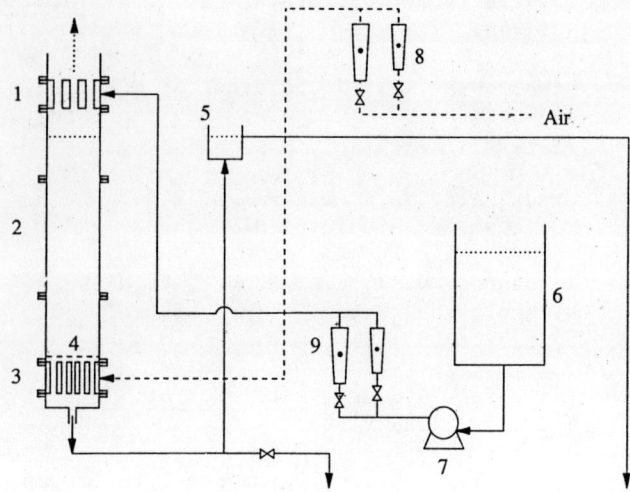

Figure 1. Experimental apparatus. 1. Liquid distributor, 2. Test section, 3. Gas distributor, 4. Retaining grid, 5. Weir, 6. Liquid reservoir, 7. Liquid pump, 8,9. Flowmeters.

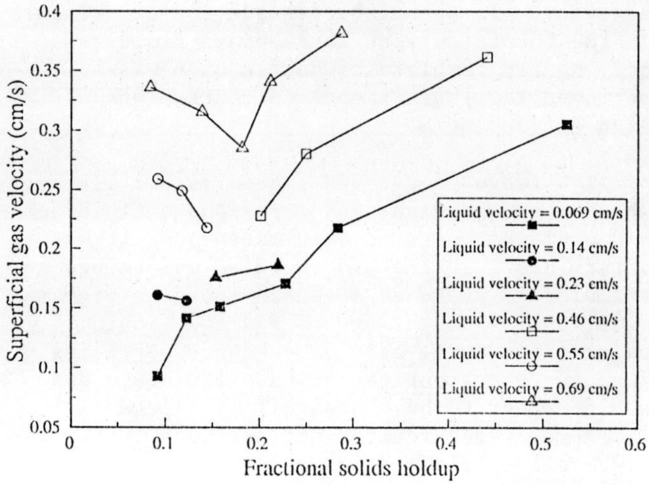

Figure 3. Superficial gas velocity at minimum fluidization vs. overall solids holdup in column for non-zero liquid velocity.

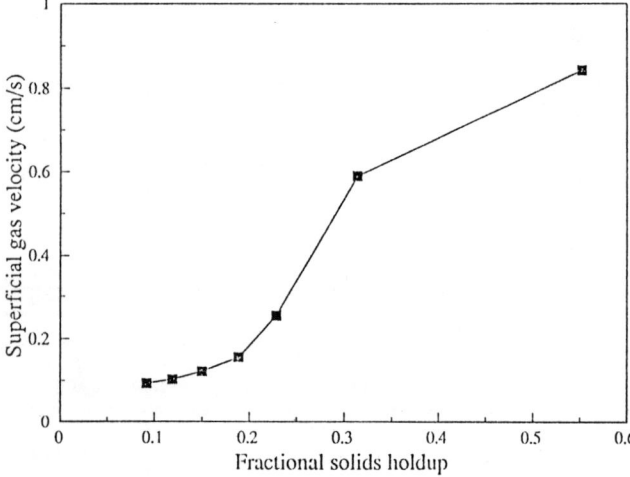

Figure 2. Superficial gas velocity at minimum fluidization vs. overall solids holdup in column at zero liquid velocity.

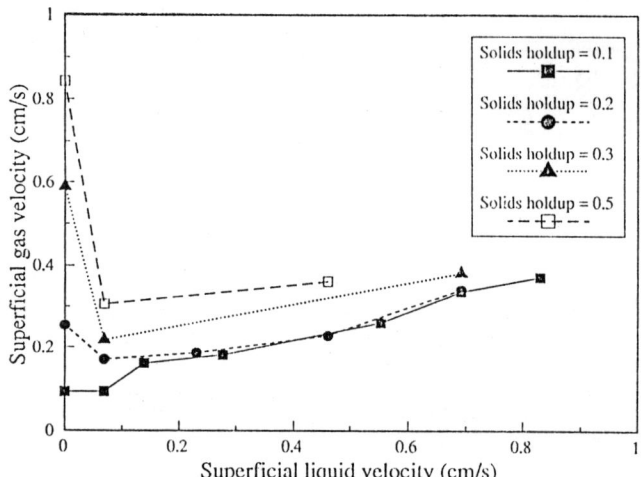

Figure 4. Superficial gas velocity at minimum fluidization vs. superficial liquid velocity at minimum fluidization.

CATALYTIC DEHYDRATION OF ETHANOL IN A FLUIDIZED BED REACTOR

P.L. Yue and N. Zarifis ■ School of Chemical Engineering, University of Bath, Bath BA2 7 AY, United Kingdom

The catalytic dehydration of ethanol over a zeolite 13X catalyst (Group A particles) has been studied in a fluidized bed reactor pilot plant. Reaction rates were measured in a complementary fixed bed reactor. The rate data were tested by several kinetic models. The variation of catalyst activity was accounted for in the kinetic model. Concentration profiles obtained from the fluidized bed reactor showed that a considerable degree of conversion occurred in the region near the distributor. A two-region model was developed to incorporate the effect of the distributor region and the channeling of bubbles further up the reactor. Plug flow is assumed in the distributor region and slug flow in the upper region. The conversion profiles given by the model show good agreement with the experimental data. Product selectivities are also well represented by the model.

It is concluded that the use of a two-region model and accurate determination of gas flow pattern will lead to a better description of fluidized bed reactors where there is some degree of slugging in the bed and the particles are of the Group A classification.

Many experimental studies designed for testing the numerous fluidized reactor models that have been developed can be found in the voluminous literature on fluidization. Reactions which lead to the formation of multiple products are particularly useful for model validation as product selectivities provide a more rigorous criterion for testing models than conversion alone. Another useful criterion for testing models is to determine how well they can predict axial concentration profiles obtained in the reactor. The above two points have been well illustrated in several studies (eg 1-4). It is also well known that distributor design and the grid region can have a significant effect on reactor performance (eg 4-7). A reliable model should therefore be able to account for the effect of the grid region.

The work reported in this paper is an extension of an earlier study by Yue and Birk (8) with a view to obtain more experimental evidence which might shed light on some of the unresolved issues. The same test reactions were conducted over the same catalyst, viz. ethanol was catalytically dehydrated to ethylene and ether over zeolite 13X in a fluidized bed reactor.

EXPERIMENTAL EQUIPMENT AND METHOD

The pilot plant fluidized bed reactor shown in Figure 1 has an overall height of 1.52 m and an internal diameter of 0.154 m. The gas distributor is a 10 mm thick stainless steel plate with multiorifices. The number and size of the orifices were chosen to ensure uniform fluidization. The stainless steel reactor has been designed to withstand pressures of up to 1600 kN m^{-2} at 773 K.

Thirteen sampling probes were installed along the the reactor for measuring the concentration profile through the bed. The lowest probe was located at 2.5 cm above the distributor for obtaining essential data at the grid region. The probes were positioned alternately along the central axis and 3.5 cm away from the reactor wall. This arrangement was designed for detecting the presence of any significant channelling.

The reactor was heated by a sheathed electrical heater coiled around the external wall. The heater was controlled by a temperature controller. Temperatures were measured by three thermocouples located at 6.5, 25.0 and 45.0 cm above the distributor and another thermocouple inserted through the catalyst loading and solid sampling port at the top of the bed. The entire reactor was fully insulated. Pressure tappings were placed immediately above and below the distributor plate.

Figure 2 gives an overview of the pilot plant. Absolute alcohol of 99.9% purity was delivered by a metering pump from two feed tanks to a heat exchanger where it was vaporised. The alcohol vapour feed was then diluted with preheated high purity nitrogen. The mixed feed was heated to reactor temperature in an electrically heated furnace which was controlled by a temperature controller. The flow was allowed to become fully developed in a wind box before entering the fluidized bed reactor. The gaseous effluent from the reactor outlet passed through a cyclone for the removal of entrained particles. All condensables were stripped by cooling before the final effluent passed into an adsorber. The adsorber was packed with 1-3 mm zeolite 13X beads. The capacity of the adsorber was sufficient to retain all unreacted reactants and products for a complete experiment. After each run the adsorber was thermally regenerated. The wall of the adsorber was also heated by a coiled electrical heater and the adsorber temperature was controlled by another temperature controller. Other safety devices of the pilot plant included pressure switches, relief valves, flammable gas detector, alarms, fire curtains and a fire-fighting system which would flood the entire working area with carbon dioxide should that be required.

Prior to each experiment, the reactor was heated overnight to the chosen temperature, with the catalyst fluidized by nitrogen, but without alcohol. At the beginning of a run, alcohol was first recycled back to the feed tanks through a bypass loop until the pumping rate became steady. The alcohol was then preheated and introduced into the feed. The reactor was allowed to reach steady state, usually in approximately two hours. For each run a complete axial concentration profile was measured, which required a minimum of six hours. Samples were taken at least in duplicate, to ensure that the results of concentrations were reproducible. These samples were analysed by a gas chromatograph under conditions identical to those used in the kinetic study.

In earlier studies of the catalytic dehydration of ethanol over zeolites (4,9,10), it was found that the rates of formation of products were affected by catalyst deactivation. Thus the activity of the catalyst must be accounted for in the modelling of reaction kinetics. The importance of accounting for catalyst activity in fluidized reactors has also been demonstrated in an independent study by Böck et al (11). In the present study, reaction kinetics have been comprehensively studied in a fixed bed reactor. The history of the catalyst activity was determined as follows. Control experiments at a pre-selected set of reference conditions were conducted in the fluidized bed reactor at random intervals following several normal runs. The conversions obtained in the control experiments at two chosen probe positions measured at steady state were used for establishing the time variation of the catalyst activity. Catalyst samples were also taken from the fluidized bed before every run. The samples were then loaded into the fixed bed reactor for the measurement of conversion under the same reaction conditions. The conversions obtained were then compared with those when a fresh catalyst was used.

In the fluidized bed reactor experiments the variables studied included reaction temperature (598 to 648 K), reactant partial pressure (0.2 and 0.4 bar) and superficial gas velocity (1.77 to 3.50 cm s^{-1}) and mass of catalyst (0 to 10 kg). The effect of sampling with the probe in the grid region at different radial positions was also examined.

RESULTS AND DISCUSSION

Rate expressions, based on two competing reactions in parallel, with ethanol dehydrating to ethylene and ether (water being the co-product in both reactions) were developed to model the rates of formation of products. The models tested included several of the power law (P-L) type and those of the Hougen-Watson (H-W) type based on Langmuir-Hinshelwood surface reaction kinetics.

In the case of ethylene formation, one P-L model and two H-W models were found to give good representation of the rate data. The reaction was irreversible with the presence of water having no effect on the formation of ethylene. Both H-W expressions assumed surface reaction as the rate controlling step but the surface reaction was monomolecular for one model and bimolecular for the other. However, in the case of ether formation, the P-L model correlated the rate data better than the H-W models. It should be noted that the P-L model was based on a reversible reaction scheme. The reverse reaction of ether hydration was found to be significant. This was because the partial pressure of water was comparable to that of ethanol, especially at higher temperatures. Thus the rate expressions chosen for use in the fluidized reactor models are as follows:

$$r_O = a_O k_O \exp(-E_O/RT) \, p_A^{n_O} \qquad (1)$$

$$r_E = a_E k_E \exp(-E_E/RT) \left\{ p_A^2 - \frac{p_E p_W}{K^*} \right\} \qquad (2)$$

It has been shown that two different kinds of active sites are involved in the dehydration reactions, one for ethylene and the other for ether (4,9). The deactivation characteristics of the two types of active sites have also been found to differ. The deactivation of the sites for ethylene formation followed the pattern of an exponential decay. A Szepe and Levenspiel (12) type of equation was developed to represent catalyst activity as a function of time.

$$a_O = 0.94 \exp(-2.76 \times 10^{-6} \, t) \qquad (3)$$

In the case of ether formation, the degree of catalyst site deactivation was negligible under the reactor conditions studied. Details and other interesting observations from the two types of deactivation experiments can be found elsewhere (13,14).

The fluidized bed reactor experiments showed that fairly high conversions to ethylene were obtained in the grid region. This observation is consistent with that reported by Yue and Birk (4). Most of the conversions were measured by the two lowest probes (2.5 and 11.0 cm), eg Figure 3.

A number of fluidized reactor models were suitably modified and tested with the experimental data. The models tested included a single-phase CSTR model, a modified Davidson-Harrison model and a modified Kato-Wen model. None of the model predictions matched the axial concentration profiles of ethylene. The CSTR model predicted much higher overall conversions to ethylene than that observed experimentally but reasonably well for ether. Both Davidson-Harrison and Kato-Wen models overpredicted the individual and total conversions although the product selectivities at the top of the bed were quite well represented.

The radial and axial concentration profiles obtained in this study show a discernible difference between the conversions measured at the center of the bed and those near the reactor wall. This difference is more pronounced at the top half of the bed, where the conversions measured with the probes near the reactor wall are higher than those measured in the center of the bed. Bearing in mind that the bulk of the reactions occurs in the dense phase, the observations suggest that there may be significant slugging in the center of the reactor. Slugging becomes more prominent at higher bed heights. Hence, the samples obtained from the centrally located probes favour the bubble phase from a height of 30 cm upwards; samples from near the reactor wall favour the dense phase. The poor performance of the reactor at higher bed heights may be attributed, at least partially, to the low interphase gas transfer associated with the large bubble slugs and the lack of contact between reactant and the catalyst particles.

Using the criterion suggested in Davidson et al (15), slugs are formed when

$$0.2 < \frac{d_b}{d_t} < 0.6 \qquad (4)$$

The bubble size distribution in the fluidized bed was evaluated by the correlation of Mori and Wen (16) which was found to best represent measurements made in the earlier study (4). The bubble diameter can reach a maximum of 8 cm in the present experiments, with the ratio of bubble to bed diameter attaining a value of 0.53. Thus slugging could well be present in a considerable part of the bed.

It is therefore proposed that the bed can be represented by a two-region model. The grid region follows a modified Davidson-Harrison model with plug flow assumed in the dense phase. The upper part of the bed is described by a Hovmand-Davidson slugging model (17) with the gas in the dense phase assumed perfectly mixed. The interphase transfer coefficient is obtained in a similar way as in the Davidson-Harrison model by adding the terms for bulk flow and diffusion.

Applying the model to the dehydration of ethanol, three component material balances of the following general form are written for the dense phase.

$$\left\{\frac{U-U_{mf}}{H'}\right\}(p_i'-p_{iH})(1-e^{-x'})$$

$$+ \frac{U_{mf}}{H'}(p_i'-p_{iD})+r_i\rho_D RT(1-e_B) = 0 \qquad (5)$$

In addition to the above, material balances are written for unit cross-section of the part of the bed where slug flow applies. The general equation is as follows:

$$\frac{U}{H'}(p_i'-p_{iH}) + r_i\rho_D RT (1-e_B) = 0 \qquad (6)$$

Equation (6) is solved for the partial pressure of ethanol in the dense phase by a Newton-Raphson scheme with p_A' taken to be that predicted by the Davidson-Harrison model at height h', the height of the lower region. Equation (7) is then solved for the partial pressures of ethanol in the dense phase. Once the partial pressures in the dense phase are known, the corresponding values in the bubble phase are calculated. The height of the lower region h' is adjusted to give the best match between the experimental value at the top of the bed and model prediction. An example of the predicted conversion profiles are compared with the experimental data in Figure 4. Both bubble and dense phase profiles are shown together with the average conversion profiles. The profiles for ether formation are not distinguishable on the graph.

There is fairly good agreement between the predicted and measured conversions throughout the bed. Experimental product selectivities are compared with model predictions in Figure 5. The bed height h' at which the transition from a bubbling to slugging bed is found to vary with superficial gas velocity and inlet partial pressure of ethanol. As ethanol partial pressure increases, with a corresponding increase in gas density, h' is raised from about 12% of the total bed height to 18%. This is consistent with what is normally observed for Group A powders when smoother fluidization and less slugging are found with increasing gas density.

CONCLUSION

Simple bubbling models, even when bubble size variation has been accounted for, cannot satisfactorily represent the data of this pilot plant study. However, a more complex model, dividing the bed into two regions, gives good agreement with the experiments. The assumption of plug flow in the dense phase with small bubbles in the grid region is appropriate, giving rise to the relatively high conversions observed. In the upper region of the bed, the assumption of slug flow accounts for the low conversions ontained.

NOTATION

- a_i dimensionless activity of catalyst for product i
- d_b bubble diameter
- d_t bed diameter
- e_B bed voidage
- E_i activation energy for the formation of i
- h' height of the lower region of the bed in the two-region model
- H' height of the upper region of the bed in the two-region model
- k_i pre-exponential factor for product i
- K^* chemical equilibrium constant for reversible ether formation reaction
- n_i order of reaction in the power law
- p_i partial pressure of i
- p_i' partial pressure of i at h'
- r_i reaction rate of formation of i
- R universal gas constant
- t time
- T absolute temperature
- U superficial gas velocity
- U_{mf} superficial gas velocity at minimum fluidization
- X' transfer parameter for the upper region of the bed
- ρ_d density of the dense phase

Subscripts:

- A   ethanol
- o   ethylene
- E   ether
- W   water

LITERATURE CITED

1. Chavarie, C. and Grace, J.R., Ind. Eng. Chem. Fundam., 14, 79 (1975)

2. Shaw, I.D., Hoffman, T.W. and Reilly, P.M., AIChE Symp. Ser. 70, (141), 41 (1974)

3. Yates, J.G. and Constans, J.A.P., Proc., Int. Symp. Fluidization and its Applications, Toulouse, 481 (1974)

4. Yue, P.L. & Birk, R.H., Chem. Eng. Res. Des., 63, 250 (1985)

5. Kolaczkowski, J.A., "Distributor design and its effect on a fluidised bed reactor", PhD Thesis, University of Bath (1981)

6. Whitehead, A.B. and Dent, D.C., Proc., Int. Symp. Fluidization, Eindhoven, 802 (1967)

7. Cooke, M.J., Harris, W., Highley, J. and Williams, D.F., Tripartite Chem. Eng. Conf., Montreal, 21 (1968)

8. Birk, R.H., "Catalytic alcohol dehydration in a fixed bed and a fluidised bed reactor", PhD Thesis, University of Bath (1983)

9. Birk, R.H., Thomas, W.J. and Yue, P.L., Chem. Eng. Res. Des., 63, 338 (1985)

10. Yue, P.L. and Birk, R.H., IChemE Symp. Ser. 87, 487 (1984)

11. Böck, W., Sitzmann, W., Emig G. and Werther, J., IChemE Symp. Ser. 87, 479 (1984)

12. Szepe, S. and Levenspiel, O., 4th Eur. Symp. Chem. React. Eng., Brussels, 265 (1971)

13. Zarifis, N., "The catalytic dehydration of ethanol in a fixed bed and a fluidised bed reactor", MSc Thesis, University of Bath (1984)

14. Yue, P.L. and Zarifis, N., World Congr. Chem. Eng. III, Tokyo, IV-128 (1986)

15. Davidson, J.F., Harrison, D., Darton, R.C. and La Nauze, R.D. in *Chemical Reactor Theory: A Review*, eds. Lapidus, L. and Amundsen, N.R., Prentice Hall, 583 (1977)

16. Mori, S. and Wen, C.Y., *Fluidization Technology I*, ed. Keairns, D.L., Hemisphere Publ. Co., 179 (1976)

17. Hovmand, S. and Davidson, J.F., Trans. Inst. Chem. Engrs., **46**, T190 (1968)

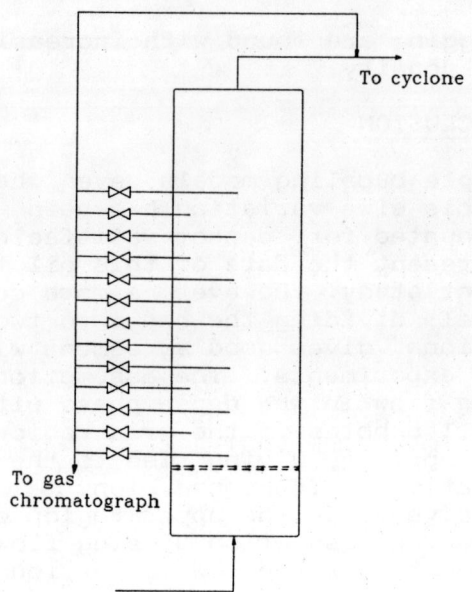

Figure 1 Fluidized bed reactor

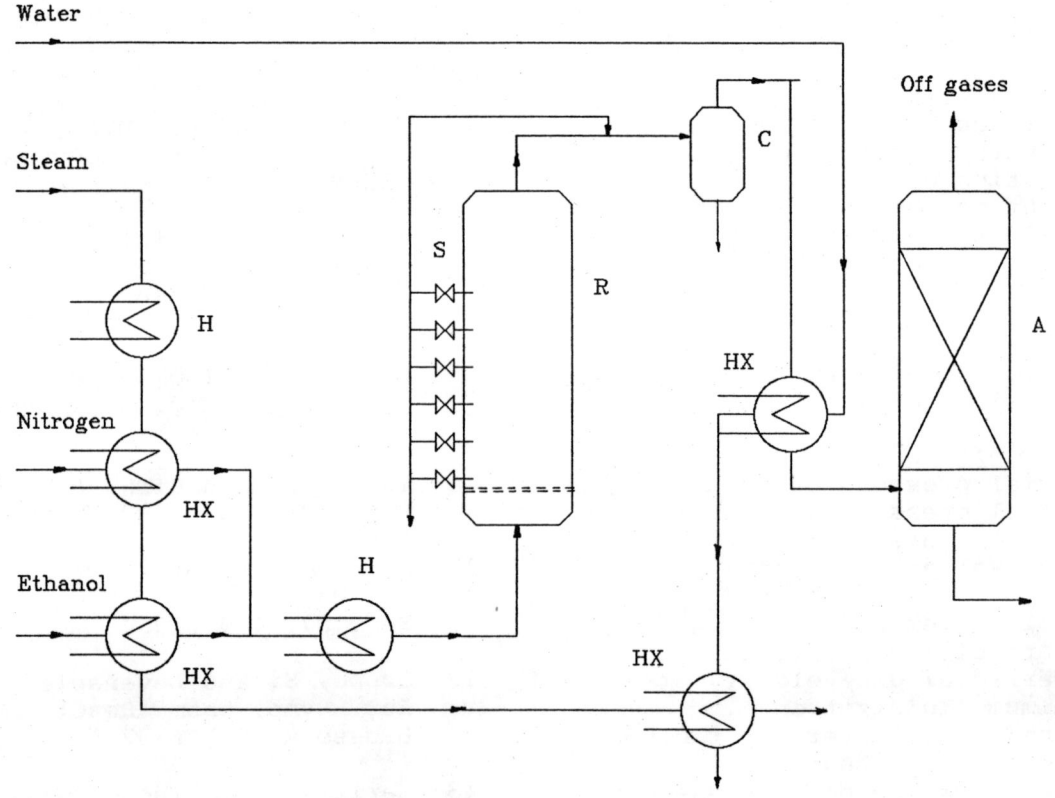

Figure 2   Schematic diagram of pilot plant

H	Heater	A	Adsorber	C	Cyclone
HX	Heat exchanger	R	Reactor	S	Sample probes

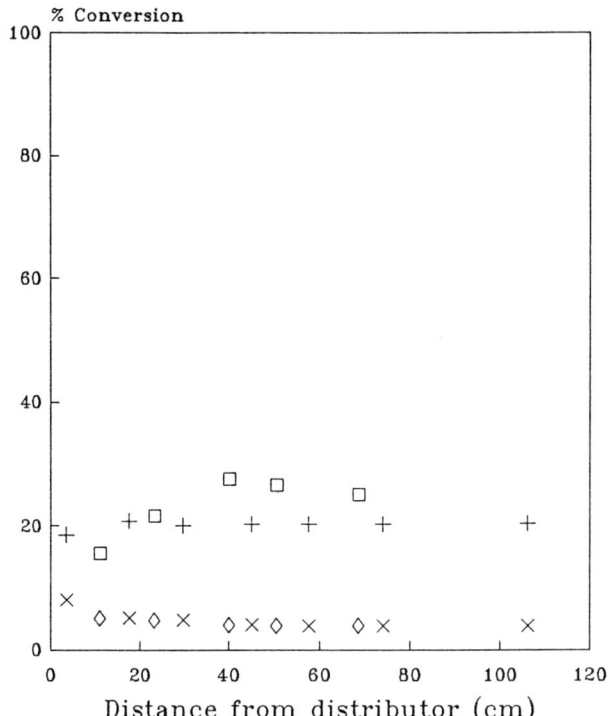

Figure 3 Axial conversion profile at 598K
  P_{AO} = 0.25 bar,
  u = 4.30 U_{mf}

+ □ Experimental data for ethylene
× ◇ Experimental data for ether
+ × Center of reactor
□ ◇ Near the reactor wall

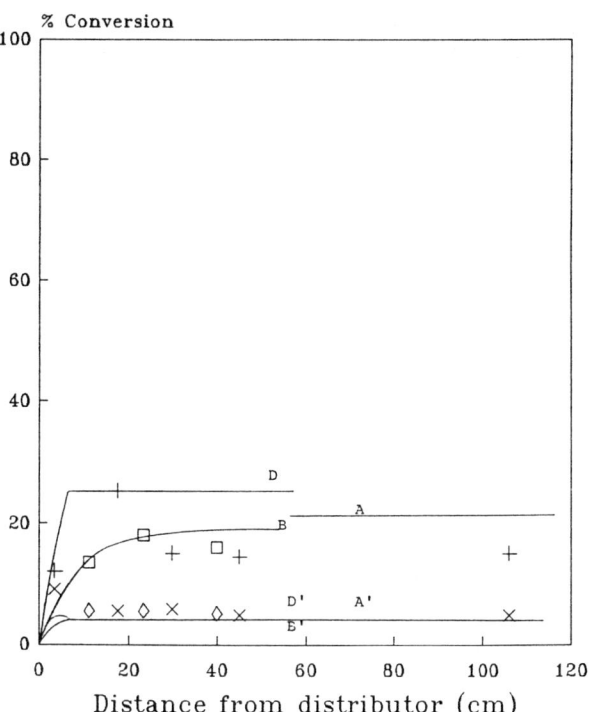

Figure 4 Comparison of conversion profiles with two-region model

+ □ Experimental data for ethylene
× ◇ Experimental data for ether
+ × Center of reactor
□ ◇ Near the reactor wall

— B   Bubble phase) theoretical
— D   Dense phase) model for
— A   Average   ) ethylene

— B' Bubble phase) theoretical
— D' Dense phase) model for
— A' Average   ) ether

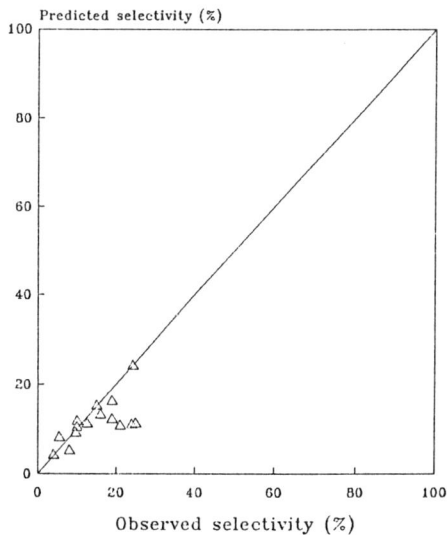

Figure 5 Comparison of selectivities with two-region model

INDEX

B
Biological phenol degradation 113
Bubble coalescence, analysis of the effects of 1
Bubble eruptions in a fluidized bed 78

C
Catalytic dehydration of ethanol 119
Cell immobilization technique 113
Cluster agglomeration formation of fine powders 72
Cohesive powder, flow conditioners for 26
Continuous depressurization of solids 61
Countercurrent three-phase fluidized bed 113

D
Dilute and dense phase gas-solids suspensions 10

E
Ethanlol, catalytic dehydration of 119

F
Fine powders, agglomeration cluster formation of 72
Flow conditioners, submicron particles as 26
Fluidized bed reactor, dehydration of ethanol in a 119
Fluidized countercurrent bed three-phase 113
Fluidized regime delineation in gas-fluidized beds 95

G
Gas distributor, design of 16
Gas motion, analysis of 78
Gas motion at the surface of 78
Gas-fluidized beds, fluidization regime delineation in 95
Gas-solid two phase flow, fine powders in 72
Gas-solids suspensions, heat transfer to 10
Group-C particles, vibro fluidization of 88

H
Heat transfer to gas-solids suspensions 10

I
Imaging experiments of a 41
Incineration of 51

L
Liquid-solid fluidized bed, effect of taper angle on 104

M
Metal capture during fluidized bed incineration 51

N
Negatively and positively charged submicron particles 26

P
Preliminary capacitance imaging experiments 41

R
Resistant gas distributor, design of 16
Restricted pipe discharge system 61

S
Solid wastes, fluidized bed incineration of 51
Solids, continuous depressurization of 61
Submicron particles, comparative evaluation of 26

T
Tapered liquid-solid fluidized bed, hydrodynamics of 104
Tube wastage, bubble coalescence on 1

V
Vertically flowing gas-solids suspensions 10
Vibro-fluidization of group-C particles 88

SYMPOSIUM SERIES

ADSORPTION

96 Developments in Physical Adsorption	230 Adsorption and Ion Exchange—'83	242 Adsorption and Ion Exchange: Recent Developments
117 Adsorption Technology	233 Adsorption and Ion Exchange—Progress and Future Prospects	264 Adsorption and Ion Exchange: Fundaments and Applications
219 Recent Advances in Adsorption and Ion Exchange		

AEROSPACE

33 Rocket and Missile Technology
52 Chemical Engineering Techniques in Aerospace

BIOENGINEERING

69 Bioengineering and Food Processing	108 Food and Bioengineering—Fundamental and Industrial Aspects	172 Food, pharmaceutical and bioengineering—1976/77
84 The Artificial Kidney	114 Advances in Bioengineering	181 Biochemical Engineering Renewable Sources of Energy and Chemical Foodstocks
86 Bioengineering ... Food	163 Water Removal Processes: Drying and Concentration of Foods and Other Materials	
93 Engineering of Unconventional Protein Production		
99 Mass Transfer in Biological Systems		

CRYOGENICS

224 Cryogenic Processes and Equipment 1982
251 Cryogenic Properties, Processes and Applications 1986

CRYSTALLIZATION

110 Factors Influencing Size Distribution	215 Nucleation, Growth and Impurity Effects in Crystallization Process Engineering	253 Fundametnal Aspects of Crystallization and Precipitation Processes
193 Design Control and Analysis of Crystallization Processes	240 Advances in Crystallization From Solutions	

DRAG REDUCTION

11 Drag Reduction
130 Drag Reduction in Polymer Solutions

ENERGY

Conversion and Transfer

5 Heat Transfer, Atlantic City	119 Commercial Power Generation	216 Processing of Energy and Metallic Minerals
57 Heat Transfer, Boston	138 Heat Transfer—Research and Design	225 Heat Transfer—Seattle 1983
59 Heat Transfer, Cleveland	162 Energy and Resource Recover from Industrial and Municipal Solid Wastes	236 Heat Transfer—Niagra Falls 1984
75 Energy Conversion Systems	174 Heat Transfer, Research and Application	245 Heat Transfer—Denver 1985
79 Heat Transfer with Phase Change	189 Heat Transfer—San Diego 1979	257 Heat Transfer—Pittsburgh 1987
87 Advances in Cryogenic Heat Transfer	202 Transport with Chemical Reactions	263 Heat Transfer—Houston 1988
113 Convective and Interfacial Heat Transfer	208 Heat Transfer—Milwaukee 1981	269 Heat Transfer—Philadelphia 1989
118 Heat Transfer—Tulsa		

Nuclear Engineering

53 Part XIII	106 Part XXII	169 Developments in Uranium Enrichment
56 Part XIV	119 Commercial Power Generation	191 Nuclear Engineering Questions Power Reprocessing, Waste, Decontamination Fusion
95 Part XX	168 Heat Transfer in Thermonuclear Power Systems	221 Recent Developments in Uranium Enrichment
104 Part XXI		

ENVIRONMENT

78 Water Reuse	165 Dispersion and Control of Atmospheric Emissions, New-Energy-Source Pollution Potential	201 Emission Control from Stationary Power Sources: Technical, Economic and Environmental Assessments
97 Water—1969	170 Intermaterials Competition in the Management of Shrinking Resources	207 The Use and Processing of Renewable Resources—Chemical Engineering Challenge of the Future
115 Important Chemical Reactions in Air Pollution Control	171 What the Filterman Needs to Know About Filtration	209 Water—1980
122 Chemical Engineering Applications of Solid Waste Treatment	175 Control and Dispersion of Air Pollutants: Emphasis on NO_X and Particulate Emissions	210 Fundamentals and Applications of Solar Energy II
124 Water—1971	177 Energy and Environmental Concerns in the Forest Products Industry	211 Research Trends in Air Pollution Control: Scrubbing, Hot Gas Clean-up, Sampling and Analysis
126 Air Pollution and its Control	184 Advances in the Utilization and Processing of Forest Products	213 Three Mile Island Cleanup
133 Forest Products and the Environment	188 Control of Emissions from Stationary Combustion Sources Pollutant Detection and Behavior in the Atmosphere	223 Advances in Production of Forest Products
137 Recent Advances in Air Pollution Control	195 The Role of Chemical Engineering in Utilizing the Nation's Forest Resources	232 Applications of Chemical Engineering in the Forest Products Industry
139 Advances In Processing and Utilization of Forest Products	196 Implications of the Clean Air Amendments of 1977 and of Energy Considerations for Air Pollution Control	239 The Impact of Energy and Environmental Concerns on Chemical Engineering in the Forest Products Industry
144 Water—1974: I. Industrial Wastewater Treatment	198 Fundamentals and Applications of Solar Energy	243 Separation of Heavy Metals and Other Trace Contaminants
145 Water—1974: II. Municipal Wastewater Treatment	200 New Process Alternatives in the Forest Products Industries	246 Advances in Process Analysis and Development in the Forest Products Industries.
146 Forest Product Residuals		265 Resource Recovery of Municipal Solid Wastes
147 Air: I. Pollution Control and Clean Energy		
148 Air: II. Control of NO_{XX} and SO_X Emissions		
149 Trace Contaminants in the Environment		
151 Water—1975		
156 Air Pollution Control and Clean Energy		
157 New Horizons for the Chemical Engineer in Pulp and Paper Technology		

FLUIDIZATION

101 Fundamental Processes in Fluidized Beds	234 Fluidization and Fluid Particle Systems: Theories and Applications	262 Fluidization Engineering: Fundamentals and Applications
105 Fluidization Fundamentals and Application	241 Fluidization and Fluid Particle Systems: Recent Advances	270 Fluidization and Fluid Particle Systems: Fundamentals and Application
116 Fluidization: Fundamental Studies Solid-Fluid Reactions, and Applications	255 New Developments in Fluidization and Fluid-Particle Systems	276 Advances in Fluidization Engineering
176 Fluidization Application to Coal Conversion Processes		
205 Recent Advances in Fluidization and Fluid-Particle Systems		

HISTORY OF CHEMICAL ENGINEERING

- 100 The History of Penicillin Production
- 235 Diamond Jubilee Historical/Review Volume

ION EXCHANGE

- 79 Adsorption and Ion Exchange Separations
- 219 Recent Advances in Adsorption and Ion Exchange
- 230 Adsorption and Ion Exchange—'83
- 233 Adsorption and Ion Exchange—Progress and Future Prospects
- 259 Recent Progress in Adsorption and Ion Exchange

KINETICS

- 25 Reaction Kinetics and Unit Operations
- 73 Kinetics and Catalysis

MINERALS

- 15 Mineral Engineering Techniques
- 85 Fossil Hydrocarbon and Mineral Processing
- 173 Fundamental Aspects of Hydrometallurgical Processes
- 180 Spinning Wire from Molten Metals
- 216 Processing of Energy and Metallic Minerals

PETROCHEMICALS

- 49 Polymer Processing
- 127 Declining Domestic Reserves—Effect on Petroleum and Petrochemical Industry
- 135 The Petroleum/Petrochemical Industry and the Ecological Challenge
- 142 Optimum Use of World Petroleum
- 212 Interfacial Phenomena in Enhanced Oil Recovery

PETROLEUM PROCESSING

- 103 C_4 Hydrocarbon Production and Distribution
- 127 Declining Domestic Reserves—Effect on Petroleum and Petrochemical Industry
- 135 The Petroleum/Petrochemical Industry and the Ecological Challenge
- 155 Oil Shale and Tar Sands
- 226 Underground Coal Gasification: The State of the Art

PHASE EQUILIBRIA

- 2 Pittsburgh and Houston
- 6 Collected Research Papers
- 88 Phase Equilibria and Gas Mixtures Properties

PROCESS DYNAMICS

- 36 Process Dynamics and Control
- 46 Process Systems Engineering
- 55 Process Control and Applied Mathematics
- 214 Selected Topics on Computer-Aided Process Design and Analysis
- 267 Process Sensing and Diagnostics

SEPARATION

- 120 Recent Advances in Separation Techniques
- 192 Recent Advances in Separation Techniques—II
- 250 Recent Advances in Separation Techniques—III

SONICS

- 109 Sonochemical Engineering

MISCELLANEOUS

- 48 Chemical Engineering Reviews
- 70 Small-Scale Equipment for Chemical Engineering Laboratories
- 112 Engineering, Chemistry, and Use of Plasma Reactors
- 125 Vacuum Technology at Low Temperatures
- 143 Standardization of Catalyst Test Methods
- 182 Biorheology
- 183 The Modern Undergraduate Laboratory Innovative Techniques
- 185 Electro Organic Synthesis Technology
- 186 Plasma Chemical Processing
- 187 Chronic Replacement of Kidney Function
- 194 Hazardous Chemical—Spills and Waterborne Transportation
- 203 A Review of AIChE's Design Institute for Physical Property Data (DIPPR) and Worldwide Affiliated Activities
- 204 Tutorial Lectures in Electrochemical Engineering and Technology
- 206 Controlled Release Systems
- 217 New Composite Materials and Technology
- 220 Uncertainty Analysis for Engineers
- 228 Problem Solving
- 229 Tutorial Lectures in Electrochemical Engineering and Technology—II
- 231 Data Base Implementation and Application
- 237 Awareness of Information Sources
- 238 New Developments in Liquid-Liquid Extractors: Selected Papers From ISEC '83
- 244 Experimental Results from the Design Institute for Physical Property Data. I: Phase Equilibria
- 247 Chemical Engineering Data Sources
- 248 Industrial Membrane Processes
- 249 Measurement of High Temperatures in Furnaces and Processes
- 252 Thin Liquid Film Phenomena
- 254 Electrochemical Engineering Applications
- 256 Experimental Results From the Design Institute for Physical Property Data: Phase Equilibria and Pure Component Properties
- 258 Fiber Optics: Processing and Applications
- 261 New Membrane Materials and Processes for Separation
- 266 Diffusion and Convection in Porous Catalysts
- 268 Membrane Reactor Technology
- 271 Experimental Results From the Design Institute for Physical Property Data: Phase Equilibria and Pure Component Properties—Part II.
- 272 Membrane Separations in Chemical Industries
- 273 Fundamentals of Resid Upgrading
- 274 Competitiveness of the U.S. Chemical Industry in International Markets
- 275 Design Institute for Physical Property Data: Ten Years of Accomplishment

MONOGRAPH SERIES

- 3 The Manufacture of Nitric Acid by the Oxidation of Ammonia—The DuPont Pressure Process by Thomas H. Chilton
- 4 Experiences and Experiments with Process Dynamics by Joel O. Hougen
- 5 Present, Past, and Future Property Estimation Techniques by Robert C. Reid
- 6 Catalysts and Reactors by James Wei
- 7 The 'Calculated' Loss-of-Coolant Accident by L.J. Ybarrondo, C.W. Solbrig, H.S. Isbin
- 8 Understanding and Conceiving Chemical Process by C. Judson King
- 9 Ecosystem Technology: Theory and Practice by Aaron J. Teller
- 10 Fundamentals of Fire and Explosion by Daniel R. Stull
- 11 Lumps, Models and Kinetics in Practice by Vern W. Weekman, Jr.
- 12 Lectures in Atmospheric Chemistry by John H. Seinfeld
- 13 Advanced Process Engineering by James R. Fair
- 14 Synfuels from Coal by Bernard S. Lee
- 15 Computer Modeling of Chemical Processes by J.D. Seader
- 16 "High-Tech" Materials by Sheldon Isakoff
- 17 Separations: New Directions for an Old Field
- 128 Biotechnology: Status and Perspectives